GUIDE TO MARINE INV

Alaska to Baja Cali

2nd Edition (Revise

by

Daniel W. Gotshall

SEA CHALLENGERS
4 Sommerset Rise, Monterey, CA 93940

2005

A SEA CHALLENGERS PUBLICATION

Copy Editor: David Behrens

Front Cover

Puget Sound King Crab	*Lopholithodes mandtil* (upper left)
Spiny Sea Star	*Poraniopsis inflata* (upper right)
Zoanthid Anemones	*Parazoanthis lucificum* (lower left)
Tochni Nudibranch	*Tochuina tetraquetra* (lower right)

Library of Congress Cataloging-in-Publication Data

Gotshall, Daniel.
Guide to marine invertebrates: Alaska to Baja California / by Daniel W. Gotshall. --2nd ed. (rev.).
p. cm.
Includes bibliographical references and index.
ISBN 0-930118-37-5: $29.95
1. Marine invertebrates--Pacific Coast (North America) 2. Marine invertebrates--Pacific Coast (North America)--Identification. I. Title.

QL 365.4.P33.G68 2005
592.17743--dc22 2005042635

SEA CHALLENGERS
4 Sommerset Rise • Monterey, CA 93940
Printed in Hong Kong through Global Interprint,Inc., Santa Rosa, CA, U.S.A.
Typography and prepress production by Diana Behrens, Danville, CA, U.S.A.

DEDICATION

This book is dedicated to Wheeler J. North; scientist, naturalist, environmentalist, pioneer in the study of marine ecology, diver extraordinaire and valued friend.

Known particularly for his work on kelp forest biology and ecology, among marine biologists Wheeler is considered a legend in his own time. My personal debt to him for his help in identifying marine invertebrates is enormous.

TABLE OF CONTENTS

ACKNOWLEDGMENTS

Neither the first nor second editions of this field guide could have been produced without the advice and assistance of the following:

For confirmation and identification of selected species: Bill Austin, sponges; Jarrett Byrnes, tunicates; Henry Chaney, bryozoans; Daphne Fautin, anemones; Bob Given, bryozoans and polychaets; Terry Gosliner, molluscs; Gorden Hendler, sea stars; F. G. Hochberg, octopus; Greg Jensen, crabs and shrimp; Gretchen Lambert, tunicates; Pam Roe, polyclad worms; Mary Wickstein, crabs; and Gary Williams, hydroids.

New geographical and depth data was supplied by Andy Lamb, David Behrens, Ryan Borema, Jay Erie, Gail Gonsalves, Dave Jones and Jong. Ann Gotshall entered the new descriptions into the computer. Kori Pennypacker typed the original manuscript and entered it into the computer.

Drawings for the pictorial key to the phyla were rendered by Dave Behrens, Doug Krause, and Helen Walker. Dan Odenweller scanned the first edition in order to provide the digital copy for editing.

I wish to thank all of the above, as well as anyone else that I may have inadvertently overlook, for their assistance and advice.

PHOTO AND ART CREDITS

Bruce Bell	160
Greg Jensen	180, back cover
Marc Chamberlain	88, 90, 91, 92, 179, 199, 200, 213, 229, 261
Dale Glantz	186
Bernie Hanby	152, 170, 178, 234, 243
Ron McPeak	113
Randy Morse	164
Alan Studley	94

All other photographs by Daniel Gotshall

"In the end we will conserve only what we love. We will love only what we understand. We will understand only what we have been taught."

-Baba Diom

INTRODUCTION

As a marine biologist for the California Department of Fish and Game for 34 years, I had the opportunity to study many species of subtidal marine invertebrates and their communities. Over the years, I observed many significant changes in some of these subtidal communities, some resulting from natural causes, others resulting from the activities of man. These changes included effects of the resurgence of the sea otter population in central California, two major El Ninos, dredging, oil spills, siltation, and sport and commercial fishing operations.

Most of the adverse impact caused by man's activities could have been prevented had those responsible been more aware of the fragility of the marine ecosystems involved. Constructive change begins with the awareness that, in most cases, what we do to one species either directly or indirectly has an effect on the other animals and plants in that community. The first step in this new understanding involves knowing something about the species composition of a particular marine community, be it a sandy bottom or a rocky reef. The next step is studying the basic survival needs of the more dominant species, i.e., feeding habits, natural mortality rates, and means of reproduction. Finally, we must know how the plants and animals interact within the community itself.

The first edition of this book was published in 1994. Since that time, we have witnessed sharp declines in the populations of all species of abalone, particularly in southern California. These declines are apparently due, for the most part, to the spread of a disease called "withering syndrome." As a result, in 1997 harvesting of all abalone south of San Francisco was prohibited.

In this edition I have added 33 species; seven sponges, one hydroid, three anemones, one polychaete worm, 12 nudibranchs, four bryozoans, one phoronid, one sea star, one brittle star and three tunicates.

The primary goal of this field guide is to help you recognize and identify those animals you observe or photograph during your dives or when tidepooling. This guide should be useful to you whether you are a biologist conducting ecological studies, a student, or perhaps just someone who is primarily interest in identifying invertebrates that you have seen or photographed. Hopefully, this ability to identify at least a few of the thousands of species of marine invertebrates that are found off our coast will allow the biologist to further his or her studies of their life histories and ecology, the student to become more familiar with animals that may become the subject of further studies, and the diver or tidepooler to better appreciate the richness of our coastal marine life communitites.

SCOPE

This field guide covers the common subtidal invertebrates that are found from Alaska to central Baja California. Animals that are usually found in depths of more than 150 feet are not included. Individual animals that are less than 1/2 inch (12 mm) in length, diameter or height, as well as cryptic animals and most planktonic invertebrates have also been omitted, with the exception of some of the larger members, such as jellyfish. For pelagic invertebrates, I recommend *Pacific Coast Pelagic Invertebrates* by Dave Wrobel and Claudia Mills. For sea slugs I recommend the new field guide *Eastern Pacific Nudibranchs - A Guide to the Opisthobranchs from Alaska to Central America* by David Behrens and Alicia Hermosillo. The best field guide for crabs and shrimps is Greg Jensen's *Pacific Coast Crabs and Shrimps* published in 1995.

This edition contains descriptions and photographs of 286 species representing 12 phyla of animals without backbones. Only a few of the hundreds of marine snails have been included because there are already a number of very complete field guides for this group. Further information is provided in the reference section. This raises an important point—this field guide should, if possible, be used in conjunction with other books on Pacific coast marine invertebrates.

SCIENTIFIC NAMES

Many people tend to be intimidated by scientific names because they are derived primarily from Greek and Latin. However, learning the scientific names of a particular plant or animal is imperative if we wish to discuss a particular species with others who may know the animal by a different common name than the one we are using. The advantage of knowing an organism's scientific name is that it is used and recognized throughout the world. Common names, by contrast, may vary from one section of the country to another. *Cancer magister,* for example, is called the dungeness crab in the North Pacific, but is called the market crab off California.

Scientific names have two parts which make up the species name: the genus and the trivial names. A genus may contain one or more species. A species is a unique interbreeding population whose members are almost identical in morphology, although color and certain other features may vary to a limited extent.

In recent years, scientists have endeavored to standardize the common names of some groups of plants and animals in order to facilitate communication with non-scientists, who are often important providers of information about various animals. For a number of years, the American Fisheries Society has been publishing a list, in book form, of both the common and scientific names of North American fishes. Approximately every 10 years this list is updated. The Society is currently undertaking the task of standard-izing the names of some of the invertebrates of North America. To date they have published lists for phyla Cnidaria, Ctenophora, and Mollusca and Class Decapoda. For more information on these lists, please see the reference section. I have used some common names from these references, as well as names coined by biologists and naturalists. In many cases, no common names exist.

HOW TO USE THIS GUIDE

When I first began observing the natural world, my primary interest was birds. Hours were spent going through my field guide looking at bird illustrations. Later, in the field, whenever I observed a bird I could not name, I usually could remember seeing its illustration in the field guide and upon returning home could go directly to the correct page. I found that not only did I learn to identify the birds, but the names tended to stay with me. Admittedly, in the beginning I was only interested in learning the common names, but it was not long before this led to an interest in learning the scientific names as well. I believe the best way to learn to identify a plant or animal is constantly to review the illustrations and descriptions in a particular guide. The best way to use this field guide is to review the drawings that make up the pictorial key on page 6 as well as the color photographs throughout the guide. When you encounter unfamiliar animals, turn to the pictorial key and try to match your animal with one of the drawings in a particular phylum. Once you find a matching phylum turn to that group and check the photos. Should you find two animals in the guide, both of which resemble the one you are trying to identify, read the size and geographic range information.

CONSERVATION

Most of the invertebrates included in this field guide are not considered edible, and therefore are not harvested by divers or tidepoolers. There is, however, a growing trade for many of these species for use in aquaria, and many divers, tidepoolers, and commercial collectors are taking animals for their private display.

At the present time, it would appear that such collecting is having a minimal impact on marine communities, as long as the collecting is not concentrated in one area and collection methods do not harm animals that are not targeted.

So far, commercial and sport harvesting of invertebrates, when managed scientifically, apparently have not adversely affected other members of the same community, at least not that biologists are aware of. Good examples of management successes are the northern California dungeness crab fishery, which harvests up to 90% of the legal-sized males in the population in some years; and the sportfishery for red abalone off Northern California. The size and bag limits for these large abalone appear to have sustained a very large and healthy population of these tasty molluscs. The opposite is true for the commercial sea urchin fishery, which began in the 1970's, peaked in the 1980's, and then declined. Hopefully, a recently developed management plan will stabilize this fishery in the future.

I firmly believe that we can continue to maintain the rich diversity and widespread abundance of our coastal invertebrates and fishery resources by using the best scientific data available in managing the harvests. This means setting aside areas where no harvest is allowed, such as underwater parks or reserves and, for each of us, it means to take from the sea only what we can and will actually use. Our marine animals and plants cannot long withstand unlimited and wasteful harvesting by any source, be it by thoughtless and greedy divers and tidepoolers or sport and commercial fishers.

NEW INFORMATION

This field guide contains both published and unpublished information; it lists observations made by myself and others. Readers of my previous books have sent me information on geographic and depth ranges and sizes of animals over the years, many of which are included here. If you observe an animal that is either farther north or farther south than any of the geographic ranges listed herein, or see one that is larger or find one that is deeper than the size and depth ranges given, I would appreciate learning of it. If possible, please include a duplicate slide or print of the animal. All relevant information that we can verify will be included in future editions and the informant will be acknowledged. Send all such information to SEA CHALLENGERS, 4 Sommerset Rise, Monterey, CA 93940.

GLOSSARY

aboral Dorsal or upper surface of sea stars and other organisims, opposite the oral surface.

antennae Paired sensory appendages on the heads of crustaceans and other organisms.

bivalve Molluscs who have two shells such as scallops, oysters, mussels and other clams.

brachial Arm or an arm-like process.

branchial plume The primary respiratory structure in nudibranchs, usually located on the rear of the back (dorsum).

carapace Hard, chitinous shell covering the bodies of crabs, shrimps, and lobsters.

chela The claw or pincher at end of anterior (frontal) appendage of crabs or shrimp.

cerata In nudibranchs these finger-like structures function as respiratory organs. They contain extensions of the digestive gland. They are located in groups along the back (dorsum).

cirri Fringe like, fleshy projections. (Plural of cirrus)

clone A single cell, or animal genetically identical to its parent, descended from one original cell or animal by fission or budding, and thus genetically identical to the original cell or animal.

commensal Two species of organisms living in intimate association, in which one member benefits, but the other is neither helped nor harmed.

copepod Small, flattened crustacean, many of which are parasitic.

coralline algae Algae that secrete calcium carbonate to form an outer skeletal material.

crustacean An animal that is a member of the arthropod class Crustacea, such as lobsters, shrimps, crabs.

cryptic Hidden, tending to conceal or camouflage by matching coloration.

detritus Mixture of material that results from a combination of rock disintegration and molecules of dead organic matter.

diatom Single cell or colonial protist whose cell walls are formed of silica; the silica remains as a skeleton after death.

distal Any point located away from center of the body; as opposed to proximal, close to the body.

encrusting Covering or forming a crust on the surface of some object.

epidodium The ridge or fold along the side of the foot of gastropods (snails).

euphausid Small, shrimp-like crustaceans, often called krill, that in some areas are a very important food source for whales.

fission A form of asexual reproduction where an animal divides into two separate, but identical animals.

flagellum A long, microscopic, whip-like external appendage of a cell, which may be used to ropel it through water.

foramen An opening or orifice.

frontal veil The frontal extension of the head on nudibranchs.

gastrovascular Functioning as both a digestive and a circulatory systems.

gill An organ used for obtaining dissolved oxygen from water.

gonad Primary sex gland, either the ovary or testis.

hermaphrodite An organism that possesses both male and female reproductive organs.

holdfast The mass of tendrils at the base of marine algae that anchors the plant to the substrate or bottom.

larvae The early form of an animal after hatching that is unlike the adult, and must pass through some form of metamorphosis to become an adult.

medusa The free-swimming, bell-shaped body of jellyfish and some hydroids.

mysid Small shrimp-like crustaceans of the order Mysidacea, that form an important food supply for fishes.

nematocyst The microscopic stinging organ in members of the phylum Cnidaria.

operculum The calcareous or chitinous cover used by snails, some worms, and other invertebrates to cover shell opening.

oral arms Fleshy lobes with frilled edges that surround the mouth on the underside of the bell of a jellyfish.

oral tenticles Sensory processes near mouth.

organic Material derived entirely from the remains or products of other living organisms.

osculum The large opening of sponges that are used to expel water.

papillae Small projections or processes on the surface of some invertebrate bodies.

parapodia Body extensions that resemble flaps on the sides of the body.

pedal laceration A means of asexual reproduction used by some anemones, where the anemone spreads out the base of the column (pedal disc) and moves on, leaving fragments which develop into anemone clones.

pedicellariae The minute, forcep-like pinchers on the outer skin of sea stars and sea urchins.

periostracum The chitinous layer that covers the shell of some molluscs.

plankton Very small, microscopic plants and animals that drift near the surface moving with the ocean currents; an important food source for larger animals.

planula The free-swimming, planktonic larvae of members of the phylum Cnidaria.

pleopod Paired abdominal appendages of crustaceans.

polyp The attached stage of hydroids, anemones, corals and gorgonians composed of a mouth surrounded by tentacles. The other end of the cylindrical body is closed and may be attached to the substrate or to other individuals in compound animals.

pore A minute opening.

protandric Hermaphroditic animals that develop and function as males, then undergo a transformation into females for the remainder of their lives.

rachis The midline structure around which the body is oriented; stem or central cord of crinoids.

radial Uniform symmetry around a central axis.

rostrum The spine-like anterior projection on the carapace of lobsters and shrimp.

sandstone Sedimentary rock composed primarily of sand.

sessile Animals permanently attached to the substrate or bottom.

siltstone Sedimentary rock composed of deposited silt or mud.

siphon Either of the two tubes of bivalve molluscs, one of which is used to transport water to the mouth or gills and the other of which is used to transport waste-laden water out of the animal.

spicule A sharp, spine-like structure; also one of the many calcareous or siliceous bodies that stiffen the body of sponges.

spongin Flexible fibers, composed of protein, that function as a skeleton in some sponges.

substrate The rocky, sandy, or other material base on which animals are attached.

symbiosis The generic term describing the behavior of animals living together in close association, including parasitism, mutualism, and commensalism.

tentacle The elongate, simple, or branched body projection in invertebrates that is used for defense, respiration, locomotion, or to acquire food.

test The calcareous internal shell of sea urchins.

tube feet The fleshy, water-filled structures of seastars, urchins, and cucumbers that are used for locomotion and for attachment to the substrate.

tubercle A small lump or projection in the skin of some invertebrates.

tunic The outer leathery covering of some sea squirts or tunicates.

valve One of the two calcareous shells of clams, scallops and piddocks.

zoanthid A colonial anemone-like cnidarian, where the polyps arise from a basal mat.

zooid A single individual in a colony of animals, particularly in bryozoans.

zooplankton The part of the planktonic community composed of animals, both larvae and adults.

PICTORIAL KEY TO PHYLA

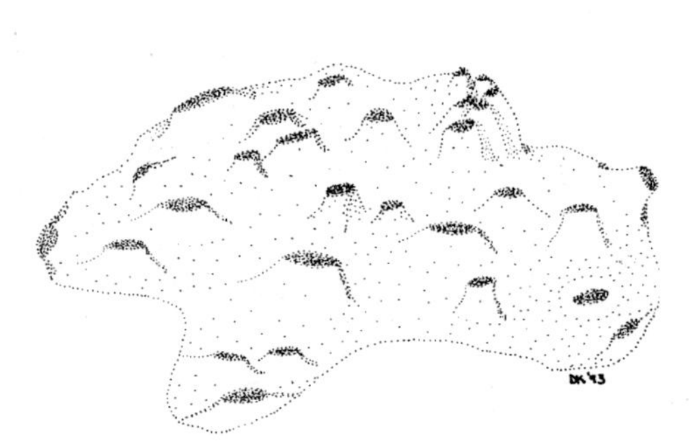

Sponges P. 11

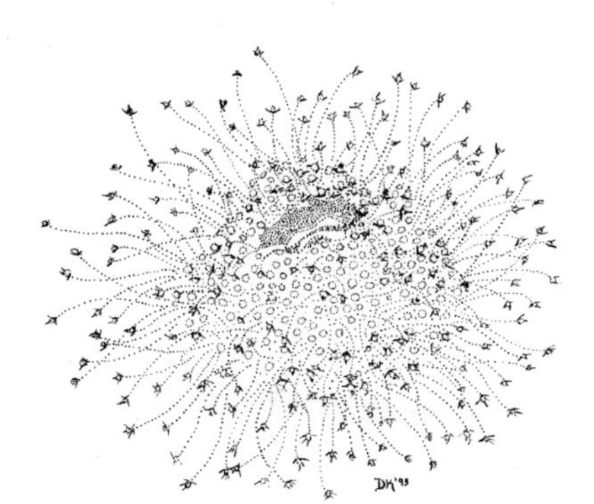

Hydroids P. 23

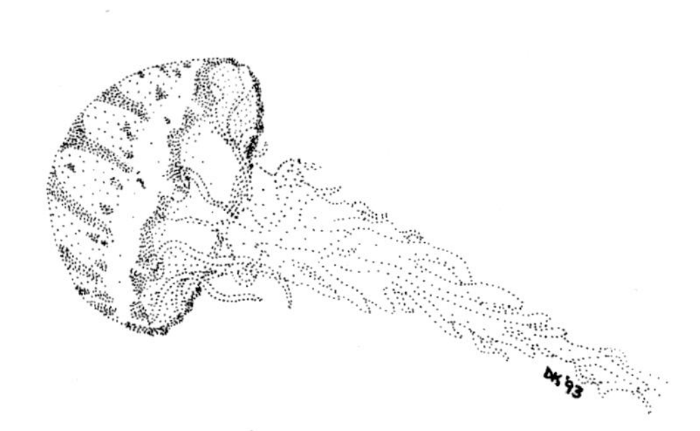

Jellyfish P. 26

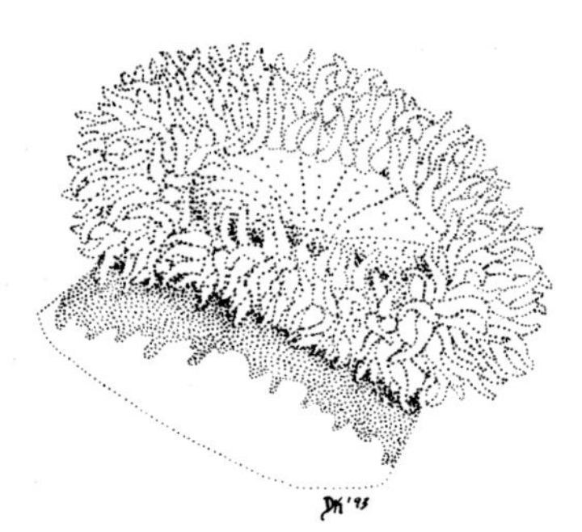

Anemones P. 28

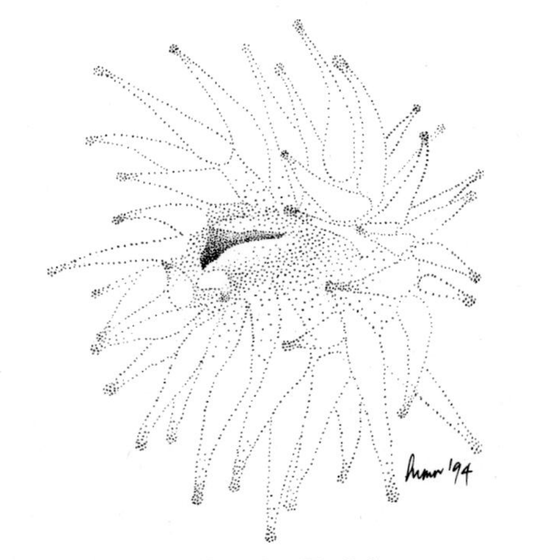

Corals P. 36

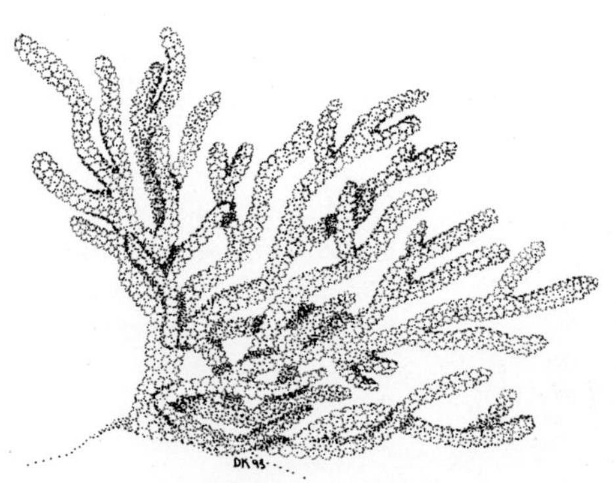

Gorgonians P. 38

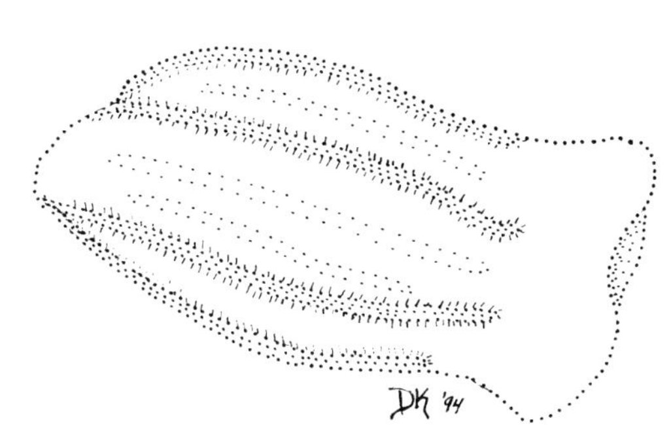

Comb Jellies P. 40

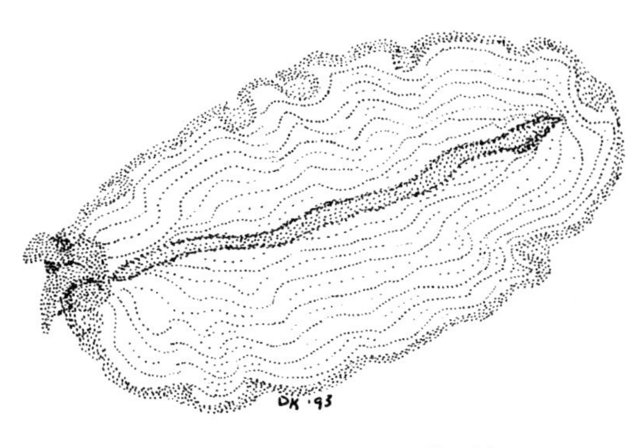

Polyclad Worms P. 41

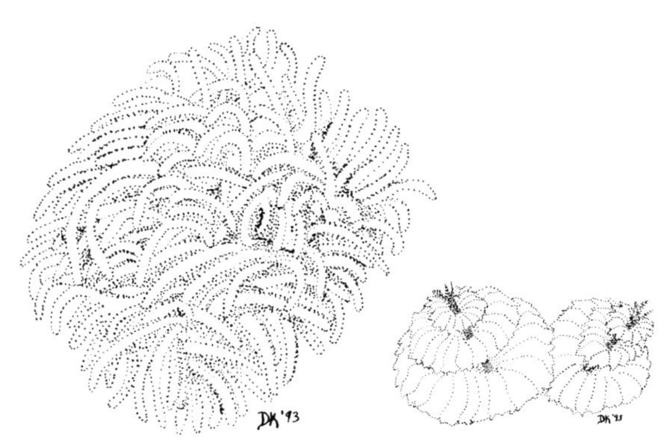

Polychaete Worms P. 43

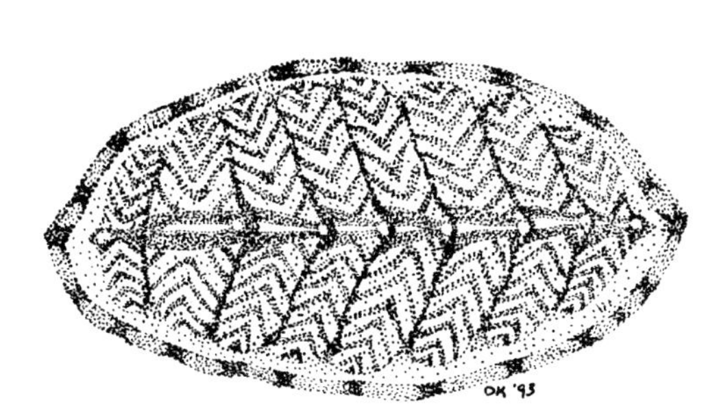

Chitons P. 48

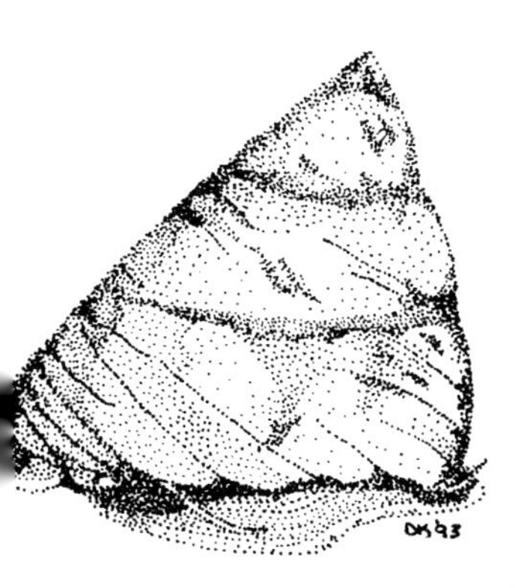

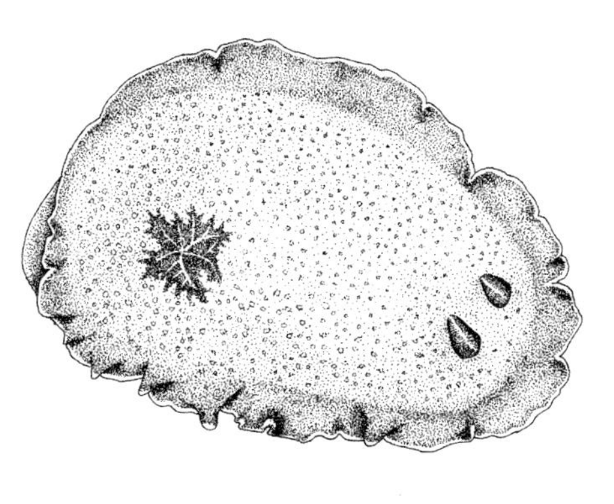

nails and Nudibranchs P. 49

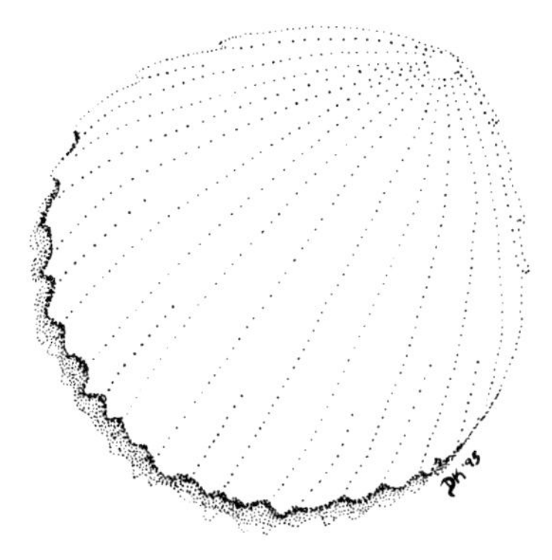

Clams and Scallops P. 61

Octopus and Squid P. 63

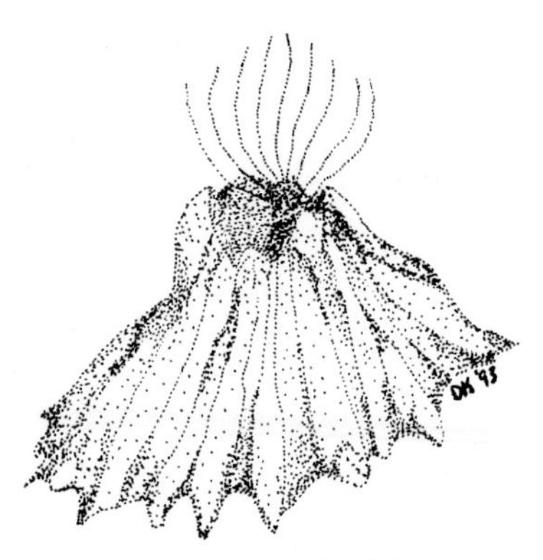

Barnacles P. 66

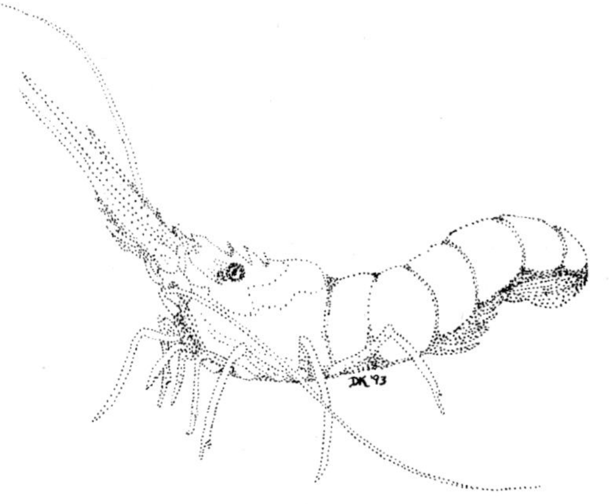

Shrimps and Lobsters P. 67

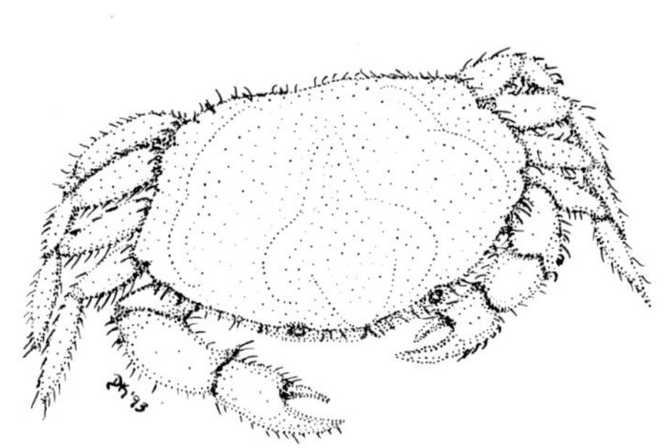

Crabs P. 70

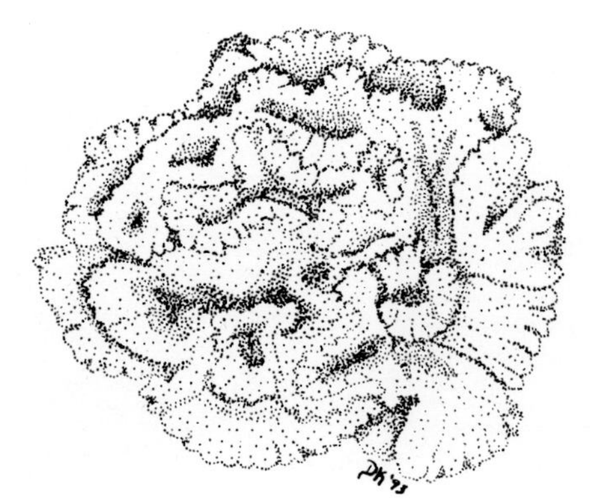

Moss Animals P. 81

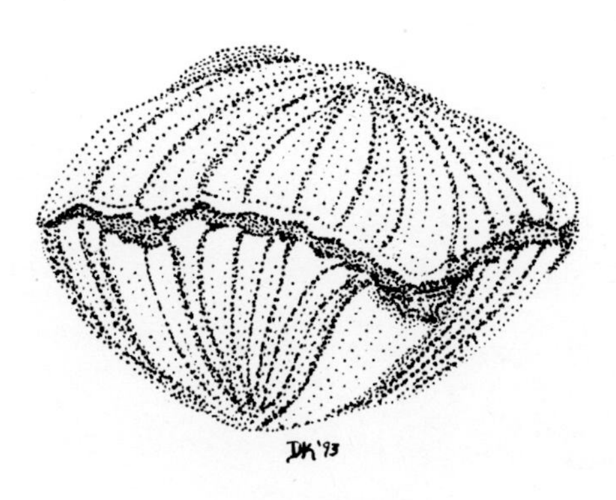

Lampshells P. 85

Phoronids P. 85

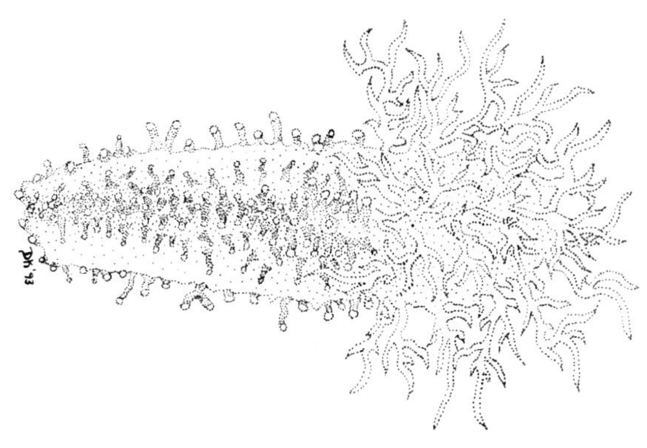

Sea Cucumbers P. 87

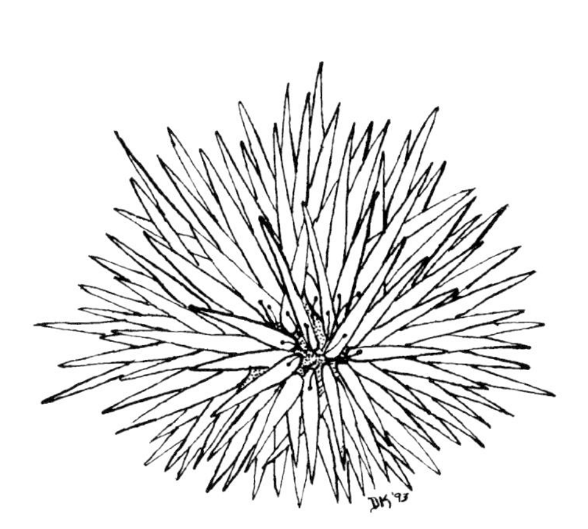

Sea Urchins P. 89

Sea Stars P. 92

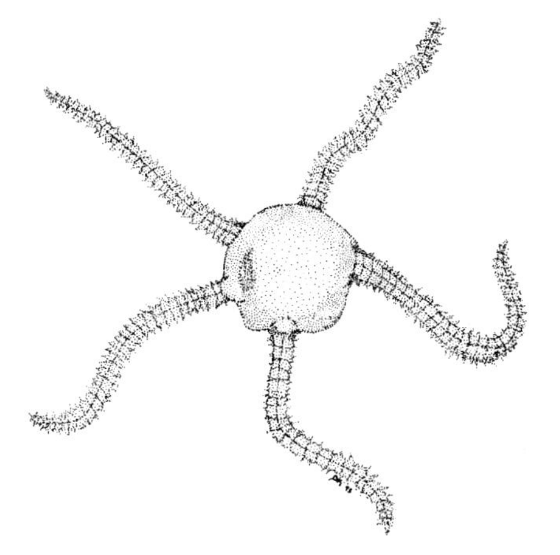

Brittle Stars P. 102

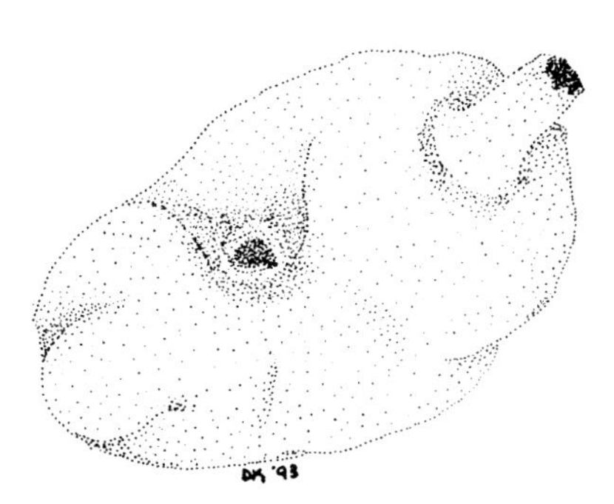

Tunicates P. 104

PHYLUM PORIFERA

Sponges

Sponges are the most primitive multicellular invertebrates. They occur in both fresh and saltwater, but the majority are found in marine waters around the world. All sponges are attached to the rocks, reefs, or other substrates where they occur. The mass of cells that form each species is organized into complex series of canals and chambers that allow water to reach each of the cells. Sponges lack tissues and organs but do have specialized cells. For example, each cell lining the chambers and canals possesses a single flagellum that is used to move water through the animal. Water enters through the myriad of small openings (or ostia) on the surface and emerges through large apertures called oscula. The movement of water through the animal provides oxygen and food, in the form of bacteria, detritus, and dissolved organic matter, and removes waste material or gametes.

Sponges reproduce by either asexual or sexual reproduction. Cloning and budding occur frequently. In sexual reproduction, the eggs are fertilized either within the animal or externally, and develop into free swimming larvae with flagella. These larvae swim about for a few days, and then settle on the bottom, where each metamorphoses into a new, tiny sponge.

Sponges are a source of food for a variety of marine animals, including some nudibranchs, chitons, sea stars, sea turtles, and tropical fishes.

Sponges produce a myriad of unique biochemical products. A considerable amount of research has been and is now being conducted on these properties. This research indicates the potential use of sponges as a source of pigments, steroids, and antibiotics as biodeterants toward predators and for use by other species in their own defense.

The taxonomy of sponges for the most part is poorly understood. Scientists use the skeletal elements (which are either calcareous or siliceous spicules or spongin fibers) to differentiate individual species. It is estimated that there are about 5000 species of sponges in the world. There are about 260 species of sponges recorded from the area between Alaska and central California.

CLASS CALCAREA
With Three Pronged Calcareous Spicules

1. SPINY VASE SPONGE
Leucandra heathi

Identification: Spiny vase sponges have a distinctive fringe of spicules around the opening (osculum) at the tip. *Size:* ranges from about 1 to 4 inches (25 to 100 mm) in height. *Range and Habitat:* Aleutian Islands south to Baja California; attached to rocks in areas of strong currents, from the intertidal to 240 ft (72 m).

2. URN SPONGE
Leucilla nuttingi

Identification: An urn-shaped, cream white sponge with a single osculum at the free end. *Size:* The average urn sponge is about 1.5 inches (40 mm) in height. *Range and Habitat:* Aleutian Islands to Baja California. Occur in groups of 5 to 10, individuals attached to rocks; low intertidal to at least 80 ft (24 m).

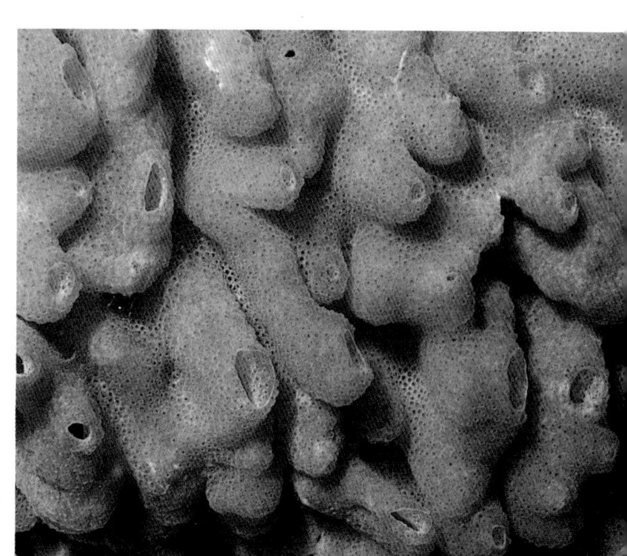

3. SPONGE
Leucetta losangelensis

Identification: This encrusting sponge has a convoluted surface with many large oscula. *Size:* To about 1 inch (25 mm) in thickness and 10 inches (250 mm) in diameter. *Range and Habitat:* Central California to Cabo San Quintin, Baja California and in the Gulf of California. Intertidal to 370 ft (111 m). *Natural History:* This and the preceding species are preyed upon by the nudibranch *Aegires albopunctatus.*

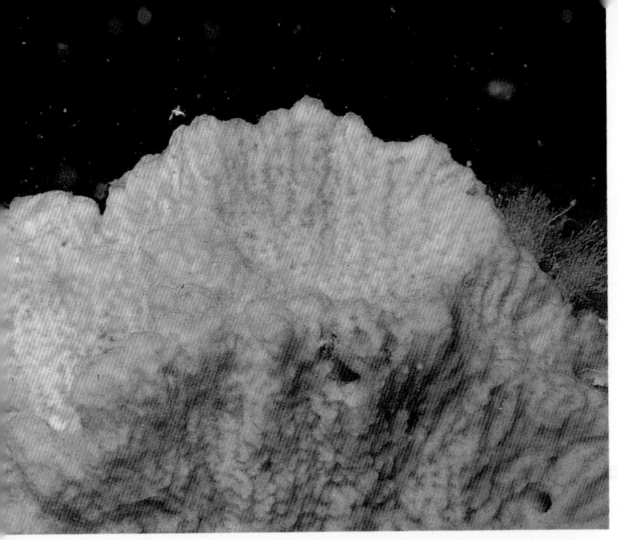

4. SPONGE

***Leucetta* sp.**

Identification: An upright, jar-shaped sponge. *Size:* Height of about 4 inches (100 mm) and width of about 8 inches (200 mm). *Range and Habitat:* Southern California, in moderately shallow water.

5. CLOUD SPONGE

Aphrocallistes vastus

Identification: A distinctive, large, white sponge. Occurs in massive formations, usually in the form of distorted vases. *Size:* Formations can reach 3 ft (90 cm) in diameter and 3 ft (90 cm) in height. *Range and Habitat:* Alaska to California, common off British Columbia; on reefs in depths of 100 ft (30 m) or more.

CLASS DEMOSPONGIAE

6. GRAY PUFFBALL SPONGE

Craniella arb

Identification: This gray-white sponge is usually ball-shaped, with a ring of spicules on the dorsal surface. *Size:* About 5 to 6 inches (125 to 150 mm) in diameter. Maximum diameter about 8 inches (200 mm). *Range and Habitat:* Strait of Georgia, British Columbia to Baja California; intertidal to at least 80 ft (24 m).

7. AGGREGATED NIPPLE SPONGE
Polymastia pacifica

Identification: The white to pale yellow, breast-shaped individuals of this aggregating sponge each have a single osculum at the tip. *Size:* Aggregations may cover an area of 2 to 3 ft (60 to 90 cm) across. Individuals 0.5 inch (12 mm) at base, 1 inch (2 cm) in height. *Range and Habitat:* Aleutian Islands to San Nicholas Island, California. On rocks with sandy pockets; very low intertidal to at least 50 ft (15 m).

8. GRAY MOON SPONGE
Spheciospongia confoederata

Identification: A massive, smooth, gray sponge (leathery in texture) with numerous crater-like oscula on the outer ridge. *Size:* This sponge can grow to a thickness of 12 inches (300 mm) and may reach 40 inches (100 cm) in length. *Range and Habitat:* Central California to central Baja California. Low intertidal zone to at least 87 ft (26 m).

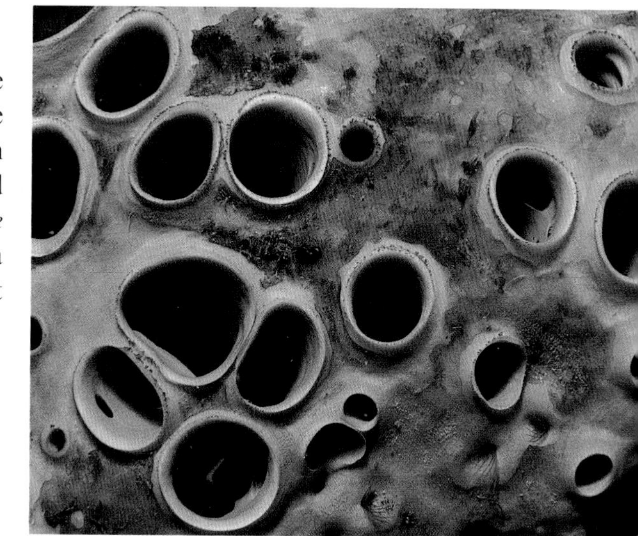

9. SPONGE
Suberites montiniger

Identification: A prominent, upright sponge with a smooth surface and many oscula. Color orange to brown. *Size:* To at least 12 inches (300 mm) across and 12 inches in height. *Range and Habitat:* On shallow reefs from the Bering Strait to British Columbia.

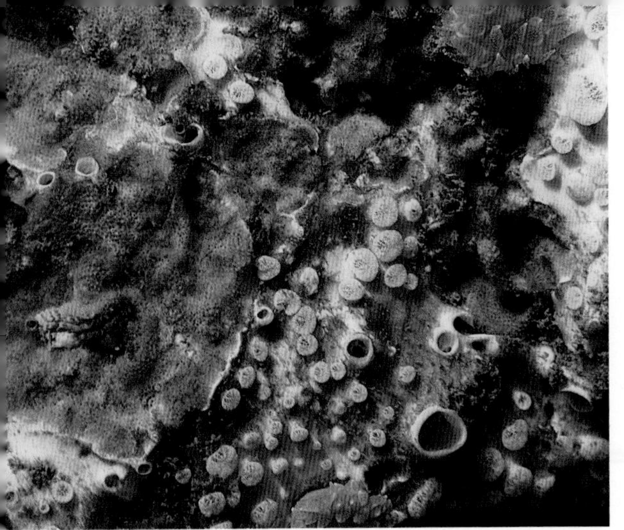

10. BORING SPONGE

Cliona celata

Identification: The exposed portion appears as yellow papillae; the rest of this sponge lies hidden in the substrate. *Size:* To at least 12 inches (300 mm) in diameter. *Range and Habitat:* Worldwide; on this coast, from Alaska to Baja California. Bores into abalone, oyster and barnacle shells. Low intertidal zone to at least 400 ft (120 m). *Natural History:* Boring is accomplished by individual cells that secrete material that fragments small pieces of the sponges substrate.

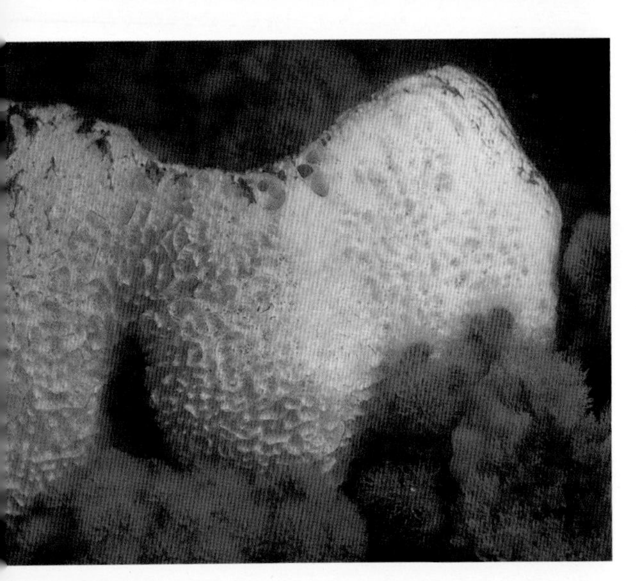

11. SPONGE

***Cliona* sp.**

Identification: A non-boring, yellow sponge forming upright masses with very rough surface texture similar in appearance to the orange puffball sponge (# 12). *Size:* Masses to 12 inches (300 mm) across and 3 inches (75 mm) in height. *Range and Habitat:* Southern Alaska to central California, on hard substrate.

12. ORANGE PUFFBALL SPONGE

Tethya aurantia

Identification: This porous, globose sponge has a very rough outer surface. The color ranges from orange to yellow. Color may be obscured by growth of commensal algae. *Size:* To about 8 inches (200 mm) in diameter. *Range and Habitat:* This cosmopolitan species occurs on our coast from southern Alaska to central Baja California, and throughout the Gulf of California. Very low intertidal to depths of 1460 ft (440 m) Usually on sides of reefs or under ledges. The blue-ring top snail (# 119) feeds on this sponge.

13. VASE SPONGE

Stylissa stipitata

Identification: A vase or goblet shaped sponge. *Size:* Diameter of open end about 2 to 3 inches (50 to 75 mm) and about 12 inches (30 cm) in height. *Range and Habitat:* Alaska to British Columbia, shallow reefs to depths of about 100 ft (30 m).

14. BREAD CRUMB SPONGE

Halichondria panicea

Identification: Usually yellow or orange to green. Surface very rough. Oscula smooth, circular, slightly raised above surface of sponge, and unevenly distributed. *Size:* Encrusting to at least 2 ft (60 cm) across and 1 inch (25 mm) thick. *Range and Habitat:* Alaska to Baja California. Intertidal to at least 300 ft (90 m) Common on deeper reefs in northern waters. *Natural History:* This sponge provides food for the cushion star (# 244) and several species of nudibranchs, including *Doris odhneri, D. montereyensis, Peltadoris nobilis, Cadlina luteomarginata,* and *Geitodoris heathi.*

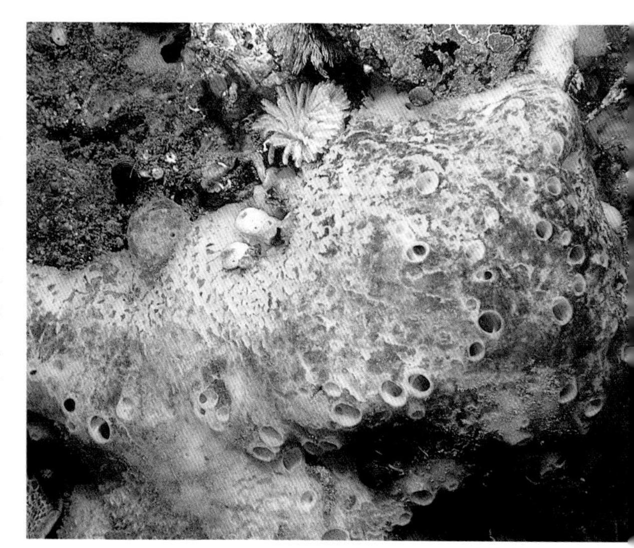

15. FINGER SPONGE

Neoesperiopsis* sp.** *(Formerly ***Isodictya)

Identification: Color ranges from red to orange. Each finger projection has a single osculum at the tip. *Size:* May reach 8 to 10 inches (200 to 250 mm) in height. *Range and Habitat:* British Columbia to northern California, in depths of 10 to at least 100 ft (3 to 30 m).

*More than one species may be involved.

16. SPONGE

Neoesperiopsis digitata

Identification: This erect, massive sponge resembles the gray moon sponge (# 8), but it is orange and the oscula are not crater-like. *Size:* Height to about 6 inches (150 mm). *Range and Habitat:* Japan to Bering Sea, south to British Columbia. On rocks in shallow, subtidal depths.

17. RED VOLCANO SPONGE

Acarnus erithacus

Identification: An encrusting sponge, scarlet to bronze colored, with numerous volcano-shaped oscula. *Size:* Masses may reach 1.5 inches (40 mm) in height and 12 inches (300 mm) in diameter. *Range and Habitat:* Southern Alaska to Gulf of California. Low intertidal zone to depths of 2333 ft (700 m). *Natural History:* The nudibranchs *Rostanga pulchra* and *Doriopsilla albopunctata* feed on this sponge.

18. SPONGE

Jones amaknakensis

Identification: An erect, rough textured sponge with many oscula on the upper surface. *Size:* To about 5 inches (125 mm) across and 3 inches (75 mm) in height. *Range and Habitat:* Bering Sea to central California, in shallow waters.

19. SPONGE

Myxilla lacunosa

Identification: This sponge has a rough surface with indistinct oscula. *Size:* To at least 5 inches (125 mm) in height. *Range and Habitat:* Aleutian Islands to British Columbia, on shallow reefs.

20. COBALT SPONGE

Hymenamphiastra cyanocrypta

Identification: An encrusting, usually cobalt blue sponge. *Size:* The thin encrusting sheets may reach 3 to 4 ft (90 to 120 cm) in diameter. *Range and Habitat:* British Columbia to Baja California; on sides of rocks or underneath ledges and in areas from low intertidal to 190 ft (57 m). *Natural History:* Symbiotic bacteria are responsible for the blue color. When the bacteria are not present, the color is usually orange.

21. SPONGE

Microciona parthena

Identification: A moderately stiff red sponge with a surface riddled with small spicules. The oscula are not readily visible. *Size:* To about 12 inches (300 mm) in diameter and 4 inches (100 mm) in height. *Range and Habitat:* Southern California and probably northern Baja California, on reefs to depths of at least 70 ft (21m).

22. RED SPONGE

Ophlitaspongia pennata

Identification: This sponge occurs in thin sheets; the oscula are small and not elevated. *Size:* Encrusting sheets up to at least 12 inches (300 mm) across and 0.5 inch (12 mm) thick. *Range and Habitat:* British Columbia to Baja California, on hard substrate; intertidal to depths of about 70 ft (21 mm). *Natural History:* The red nudibranch, *Rostanga pulchra,* lays its eggs on and feeds on this sponge.

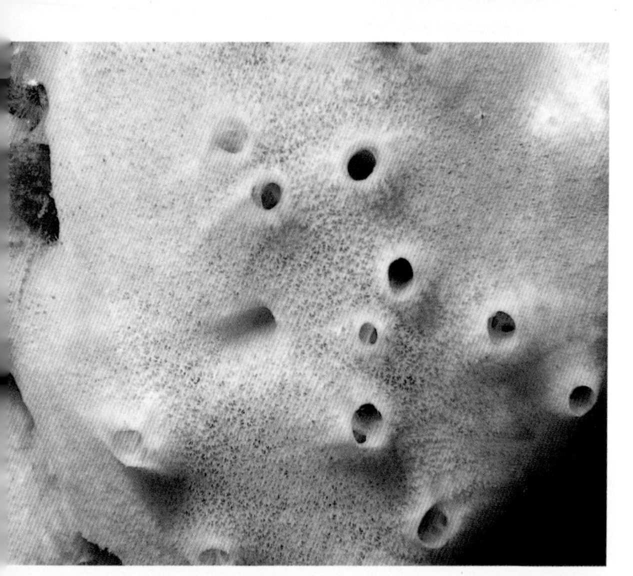

23. SPONGE

***Haliclona* sp.**

Identification: A white to tan encrusting sponge, with prominent protruding oscula. *Size:* Occurs in sheets up to 2 ft (60 cm) in diameter and 1 inch (25 mm) in height. *Range and Habitat:* Common on central California reefs in depths less than 100 ft (30 m).

24. SPONGE

Haliclona permollis

Identification: An encrusting sponge with a pitted surface and many oscula. *Size:* Up to at least 6 inches (150 mm) across and 1 inch (25 mm) in height. *Range and Habitat:* British Columbia to southern California on intertidal and shallow subtidal reefs.

25. SPONGE

Sigmadocia edaphus

Identification: This is a very rough textured sponge that forms erect tubes with a single osculum at the tip. *Size:* Base of sponge at least 6 inches (150 mm) across and 3 inches (75 mm) in height. *Range and Habitat:* British Columbia to Gulf of California, on reefs in littoral waters.

26. WHITE FINGER SPONGE

Toxadocia* spp.

Identification: A whitish sponge that lacks large oscula and forms masses of finger-like projections. *Size:* A single "finger" may reach 12 inches (300 mm) in length and 1 inch (25 mm) in diameter. *Range and Habitat:* The species illustrated has been observed off central and southern California. Common in Carmel Bay, California from 40 to at least 100 ft (12 to 30 m).

*More than one species may be involved.

27. ORANGE SPONGE

Cyamon argon

Identification: A firm, encrusting, wrinkled orange to red sponge. *Size:* To at least 12 inches (300 mm) across and 4 inches (100 mm) in height. *Range and Habitat:* Central California to Baja California on reefs in depth of 20 to 50 ft (6 to 15 m).

28. SULPHUR SPONGE

Aplysina fistularis (Formerly ***Verongia aurea)***

Identification: Usually bright yellow with an irregular surface and a few large oscula, sometimes at tips of projections. *Size:* Encrusting; 2 to 3 inches (50 to 75 mm) deep and 5 to 8 inches (125 to 200 mm) in width. *Range and Habitat:* Cosmopolitan; on this coast, southern California to Gulf of California. Found on rocks from intertidal zone to at least 100 ft (30 m). *Natural History:* The shelled notaspidean opisthobranch, *Tylodina fungina,* feeds on this sponge.

29. CHIMNEY SPONGE

Rhabdocalyptus dawsoni

Identification: This large, long, round sponge has a large osculum with a white edge on the interior. The exterior of this sponge is covered with bristles. *Size*: Height to about 5 feet (1.5 m). *Range and Habitat*: Alaska to southern California, on slopes of reefs and rocks to depths of at least 100 ft. (36 m).

30. SPINY TENNIS BALL SPONGE

Craniella spinosa

Identification: This ball-shaped sponge has one or more large oscula and is covered with spines. *Size*: Height to about 3 inches (78 mm). *Range and Habitat*: Aleution Islands south to at least British Columbia.

31. PEACHBALL SPONGE

Suberites montiniger

Identification: This irregular round sponge has several medium-sized oscula. The color ranges from brown to pink or apricot. *Size*: Width to about 8 inches (20 cm). *Range and Habitat*: British Columbia to Baja California, subtidal to about 100 ft. (30 cm). Usually occurs in high current areas.

32. SMOOTH SCALLOP SPONGE

Mycale abhaerens

Identification: This encrusting sponge is usually brown, grey or violet. Grows on scallop shells. If torn open, the prominent thick fibers are notable. *Size*: Thickness to about 1/2 inch (10 mm). *Range and Habitat*: Bering Sea to Washington, on scallop shells, particularly *Chlamys* spp. *Natural History*: The sea lemon #145 feeds on members of this genus.

33. ROUGH SCALLOP SPONGE

Myxilla incrustans

Identification: This encrusting sponge also grows on scallops; it differs from #32 in its thickness, and with a more prominent, lumpy oscula. The color ranges from golden yellow to light brown. *Size*: Thickness to about 1/2 inch (10 mm). R*ange and Habitat*: Arctic Ocean to southern California on scallop shells. *Natural History*: Several species of nudibranch feed on this sponge, including *Cadlina luteomarginata*.

34. SPONGE
Iophon chelifer

Identification: This white to yellow sponge forms encrusting masses as well as bush-like, erect growths. The branches of the erect form are in the shape of small tubes. *Size*: Height to about 12 inches (30 cm) and width to about 3 ft. (1 m). *Range and Habitat*: Alaska to southern California in depths of about 40 ft. (10 m) to at least 100 ft. (36 m). *Natural History*: The wrinkled sea star (#243) feeds on this sponge.

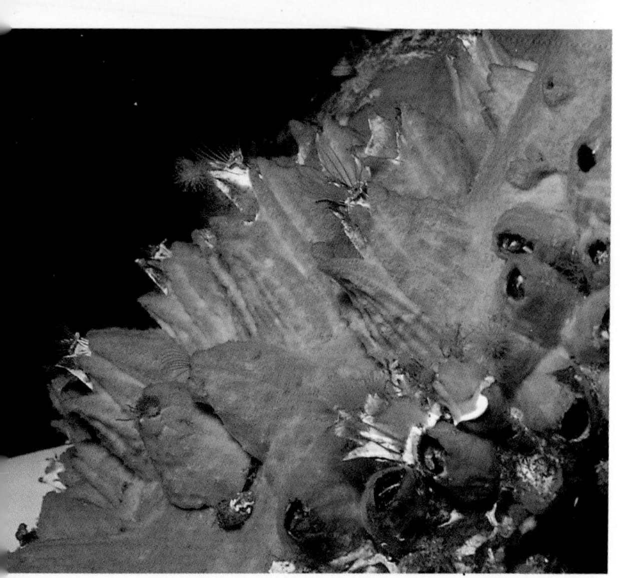

35. SPONGE
Podotuberculum hoffmani

Identification: The thin encrustations can be found on barnacles and other subtrates. *Size*: Thickness to about 1/8 inch (3 m). *Range and Habitat*: British Columbia to Washington. Intertidal to shallow subtital depths.

PHYLUM CNIDARIA
Hydroids, Anemones, Jellyfish, Corals and Gorgonians

This very diverse group of radially symmetrical animals is found in all of the world's oceans. A few species also occur in freshwater. These animals inhabit most of the marine habitats from the intertidal to the deepest oceans.

The cnidarians include mobile as well as sessile forms. Sessile animals are called polyps and mobile forms are referred to as medusae. There are two types of polyps in most sessile species, feeding and reproductive.

Cnidarians are an ancient group. Their fossil remains have been found in Cambrian rocks (over 500 million years ago). The basic anatomy of a typical cnidarian consists of a body containing the gut and a mouth usually surrounded by tentacles. The tentacles are loaded with stinging capsules called nematocysts. The nematocysts are used to capture food and for defense. When the capsule explodes, a hollow thread is released that penetrates the prey or predator. All members of this group are omnivores. Reproduction is accomplished sexually as well as asexually. Fertilization of the egg can take place within the animal or in the open water. Asexual reproduction is accomplished by budding or fission.

Approximately 350 species of hydroids, 20 scyphozoans, and 123 species of anthozoans have been recorded on this coast.

CLASS HYDROZOA
Hydroids

36. SEA FERN HYDROID
***Abietinaria* sp.**

Identification: This fern-like hydroid has a main stalk with side branches all on one plane. The feeding polyps alternate on opposiste sides of the branches and stalk. The color varies from buff to orange. *Size*: Height of colony to about 6 inches (15 cm). *Range and Habitat*: Alaska to San Diego, California; on rocks, from low intertidal to shallow subtidal.

37. PINK-MOUTH HYDROID
Tubularia crocea

Identification: Occurs in dense clusters. The polyps are pink to red, branched stalks are brown, each polyp has about 20 tentacles. *Size:* To about 6 inches (150 mm) in height. *Range and Habitat:* Circumpolar; on this coast, Gulf of Alaska to southern California. In shallow waters, on dock pilings, rocks, and boat bottoms. *Natural History:* Eggs are fertilized internally by free-swimming sperm. Feeds on zooplankton. This hydroid is preyed upon by several species of aeolid nudibranchs.

38. ORANGE HYDROID
Garveia annulata

Identification: This distinctive orange hydroid occurs in clusters of 20 to 30 stems and polyps. *Size:* Height to 6 inches (150 mm). *Range and Habitat:* Alaska to southern California. Low intertidal rocks to depths of at least 400 ft (120 m). *Natural History:* Seasonally abundant. Colonies are either male or female. Eggs are fertilized internally by free swimming sperm.

39. HEDGEHOG HYDROID

Hydractinia milleri

Identification: Colony consists of a mass of pinkish, stalked polyps rising from a basal mat. *Size:* Height to about 1 inch (25 mm). Width of base to about 3 inches (75 mm). *Range and Habitat:* British Columbia to southern California. On sides of rocks and other substrates from low intertidal to depths of at least 80 ft (24 m).

40. BELL MEDUSA

Polyorchis penicillatus

Identification: A bell-shaped, transparent medusa with a single core of tenacles and red spots on lower margin of the bell. *Size:* To about 4 inches (100 mm) in length. *Range and Habitat:* British Columbia to San Diego, California. Planktonic; occasionally rests on bottom in nearshore waters. *Natural History:* Feeds on other smaller members of plankton as well as small crustaceans. They mature as they swim through eel grass beds. Sexes are separate. Apparently there is no polyp stage; the medusa develops from a planula larvae.

41. OSTRICH-PLUME HYDROID

Aglaophenia struthionides

Identification: The large, feather-like brown plumes consist of a central stalk with numerous branches, in one plane. *Size:* Plumes to at least 5 inches (125 mm) tall. *Range and Habitat:* Alaska to Baja California, on hard substrate. Rocky intertidal to depths of 500 ft (160 m). *Natural History:* The colonies feed on small plankton that pass within reach. Sexes are separate. A free-swimming medusa stage does not occur. Eggs are retained by the females and develop into worm-like planula that eventually develop into new colonies.

42. HYDROID

***Plumularia* sp. 1**

Identification: The colony consists of upright plumes that resemble feathers. The polyps are widely spaced on the branches. *Size:* Height of colony to about 2 inches (50 mm). *Range and Habitat:* Several species are found from Alaska to Baja California; intertidal and sublittoral.

43. HYDROID

***Plumularia* sp. 2**

Identification: A colony consisting of feather-shaped plumes. *Size:* Stalks to 5 inches (125 mm) in height. *Range*: Members of this genus range from Alaska to Baja California.

44. CALIFORNIA HYDROCORAL

Stylaster californicus (Formerly ***Allopora***)

Identification: The branching colonies formed by a calcareous skeleton are often confused with true corals. The colors of the colonies range from pink to a dark blue. At least three other species of hydrocoral that are similar in appearance occur off British Columbia and Alaska. *Size:* Colonies to 12 inches (300 mm) in height. Width of base may reach 24 inches (60 cm). *Range and Habitat:* Northern California to San Benitos Islands, Baja California. In depths of about 40 to 175 ft (12 to 55 m). *Natural History:* Studies have shown that it requires 20 or more years for a colony to grow to heights of 30 cm.

45. ENCRUSTING HYDROCORAL

Stylantheca porphyra

Identification: These colonies form flattened encrusting sheets sometimes with protrusions about 1 inch (25 mm) in height or less. The color ranges from dark reddish to pink to reddish purple. At least one other species occurs off British Columbia. *Size:* Sheets may reach 3 ft (90 cm) in width. *Range and Habitat:* British Columbia to southern California, low intertidal to depths of less than 100 ft (30 m). *Natural History:* Each pit on the surface of the colony contains 1 to 12 feeding polyps.

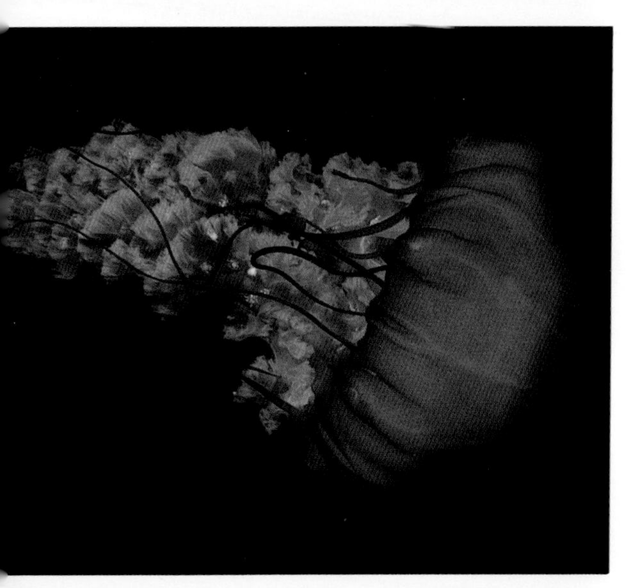

CLASS SCYPHOZOA

Jellyfish

46. BROWN JELLYFISH

Chrysaora fuscesens

Identification: The yellow-brown medusa (bell) with 24 dark brown tentacles and yellowish oval arms are very distinctive. *Size:* The bell may reach 12 inches (30 cm) in diameter. The entire animal can reach 6 to 8 ft (1.8 to 2.4 m) or more. *Range and Habitat:* Alaska to southern California; pelagic. Usually occurs off central California in the winter. *Natural History:* This jellyfish possesses very potent stinging cells. The tentacles, in particular, should be avoided.

47. PURPLE JELLYFISH

Chrysaora colorata

Identification: The bell (medusa) of this jellyfish is usually silver-white, with purple bands radiating from the center. There are also purple spots on most individuals. *Size:* The bell may reach 30 or more inches (80 cm) in diameter. The length of the individual can reach 6 to 8 ft. (1.8 to 2.4 m) or more. *Range and Habitat:* Central California and south; found in most of the oceans of the world. Pelagic; occurs inshore in our area in late fall and winter.

48. MOON JELLYFISH

Aurelia aurita

Identification: Translucent jelly with a dish-shaped bell. The tentacles and oral arms are very short. Four horseshoe-shaped gonads, visible through the bell, are distinctive. *Size:* Bell diameter to 16 inches (40 cm). *Range and Habitat:* Worldwide in temperate seas. On this coast, from Alaska to southern California; pelagic. *Natural History:* The eggs are fertilized internally by sperm released by the male. Embryos develop on grooves in the oral arms.

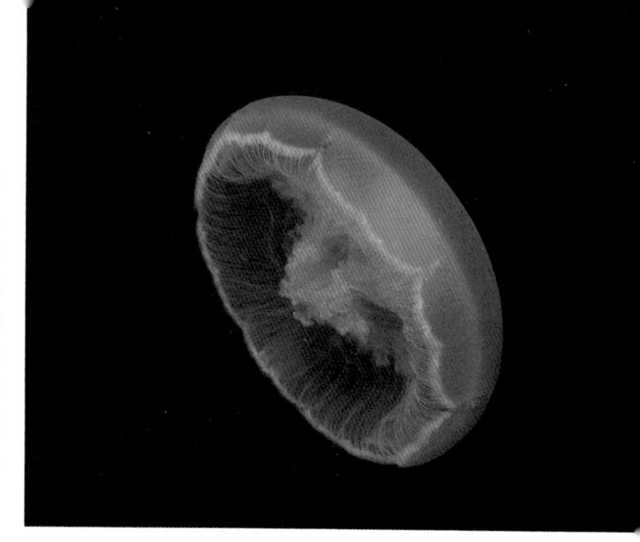

49. LION'S MANE JELLYFISH

Cyanea capillata

Identification: The bell of this brownish jellyfish is flattened, with a thick center and thinner margin. The margin consists of eight lobes. There are several hundred tentacles. *Size:* Bell diameter to 20 inches (51 cm). Tentacle length to at least 30 ft (9 m). *Range and Habitat:* Atlantic and Pacific oceans. On this coast, Alaska to southern California; pelagic. *Natural History:* This jellyfish has potent stinging cells and should be avoided.

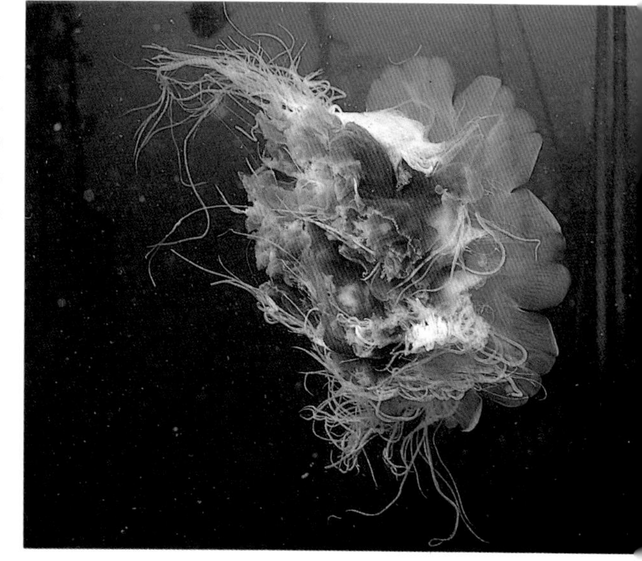

50. JELLYFISH

Phacellophora camschatica

Identification: This jellyfish is often confused with the Lion's Mane Jellyfish (# 49). It can be identified by a transparent margin consisting of 16 large lobes that alternate with smaller lobes. Each large lobe carries a cluster of up to 25 tentacles. *Size:* Bell width to 24 inches (61 cm). *Range and Habitat:* Worldwide in temperate oceans; on this coast, from Alaska to southern California. Pelagic.

CLASS ANTHOZOA
Anemones, Corals, and Sea Pens

51. WHITE-SPOTTED ROSE ANEMONE

Urticina lofotensis (Formerly ***Tealia lofotensis***)
Identification: The column is a very distinctive red, with white spots. The tentacles are scarlet to crimson and lack bands. The smooth column lacks tubercles. *Size:* Column diameter to about 4 inches (100 mm). *Range and Habitat:* Circumpolar; on this coast, Alaska to southern California (Channel Islands National Park). Attached to rocks, pilings, and marina floats. Low intertidal to about 75 ft (25 m).

52. FISH-EATING URTICINA

Urticina piscivora (Formerly ***Tealia piscivora***)
Identification: This anemone has a deep, red column. The tentacles are usually white, but occasionally are red. *Size:* Tentacular crown to 8 inches (200 mm) in diameter. *Range and Habitat:* Alaska to La Jolla, California, on sides of rocks from the low intertidal to about 160 ft (48 m). *Natural History:* Feeds on fish as well as a variety of invertebrates.

53. PAINTED URTICINA

Urticina crassicornis (Formerly ***Tealia***)
Identification: The slender tentacles of this anemone have a blunt, bulbous tip and are white with a red band. The column is usually a mottled red and olive. *Size:* Tentacular crown to about 8 inches (200 mm) in diameter. *Range and Habitat:* Alaska to southern California, uncommon in California. On rocks; intertidal to about 100 ft (30 m). *Natural History:* Feeds on crabs, sea urchins, molluscs, and fish.

54. STUBBY ROSE ANEMONE

Urticina coriacea

Identification: The tentacles on this anemone are short, stout, and blunt; tentacle colors include pink, green, and blue with one or more white bands. Column usually buried in sand. *Size:* Column diameter to 4 inches (100 mm); tentacular crown diameter to about 6 inches (150 mm). *Range and Habitat:* Alaska to southern California. Attached to rocks, but usually buried partially in sand or shell debris. Depths from 40 to 150 ft (12 to 45 m). *Natural History:* The leather star, *Dermasterias imbricata,* (# 238) preys on this anemone.

55. SAND-ROSE ANEMONE

Urticina columbiana

Identification: These large, sand dwelling anemones have large, white tentacles and a column cov-ered with tubercles that have attached sand and shell fragments. *Size:* Diameter of tentacular crown to about 14 inches (35 cm). *Range and Habitat:* British Columbia to Baja California. Buried in sand and mud bottoms. In depths of about 40 to 150 ft (12 to 45m).

56. MC PEAKS URTICINA

Urticina mcpeaki

Identification: A bright red anemone with wart-like, sticky tubercles on the column, usually encrusted with debris. These tubercles are arranged in longitudinal rows. The long, slender tentacles (when expanded) are encircled with a light colored band at their base. Differs from #51 in having an oral disc with radial banding. *Size*: Diameter to about 3 inches (7.5 cm). *Range and Habitat*: Santa Barbara, California to northern Baja California. Subtidal depths to about 100 ft. (36 m). Attached to rocks usually covered with debris such as sand and silt.

57. PROLIFERATING ANEMONE

Epiactis prolifera

Identification: This anemone has a low, squat column, with a broad base with 48 to 96 short conical tentacles. Each tentacle has a pore at the tip. The color is highly variable. *Size:* Tentacular crown to about 2 inches (50 mm) in diameter. *Range and Habitat:* Alaska to southern California. Attached to rocks, eel grass, and kelp; mid-intertidal to depths of about 50 feet (15 m). *Natural History:* Reproduce sexually. Most of the young adults are females, but they also have testes that develop along with ovaries. Sperm is produced only by large, older females.

58. CRIMSON ANEMONE

Cribrinopsis fernaldi

Identification: Distinctive crimson-colored, very long tentacles, with wart-like projections that contain the stinging cells. *Size:* Column diameter to about 8 inches (200 mm). Height to about 10 inches (250 mm). *Range and Habitat:* Gulf of Alaska to Washington, attached to rocks in subtidal depths to about 1000 ft (300 m).

59. GIANT GREEN ANEMONE

Anthopleura xanthogrammica

Identification: The column of this large, solitary anemone is covered with adhesive tentacles. The abundant tentacles are short, conical, and either pointed or blunt. *Size:* Column diameter to about 7 inches (175mm). Height to 12 inches (300 *mm). Range and Habitat:* Alaska to central Baja California. On rocks; low intertidal to about 100 ft (30 m). *Natural History:* The brilliant green color is due to both the green pigment in the outer skin and symbiotic zooxanthellae that live in cells lining the gut. The anemone farms some nutrition from the organic material synthesized by the zooxanthellae.

60. AGGREGATING ANEMONE

Anthopleura elegantissima

Identfication: These anemones occur in aggregations. The column has longitudinal rows of adhesive tubercles and the tentacles are short and abundant. *Size*: Column diameter to about 2.5 inches (60 mm) and tentacular crown diameter to about 3.5 inches (80 mm). *Range and Habitat*: Alaska to Baja California. Attached to rocks, pier pilings, or other manmade structures. Intertidal to shallow subtidal.

61. GREEN ANEMONE

Anthopleura sola

Identification: These solitary anemones are similar to the intertidel Aggregating anemones, *A, elegantissima,* but the tentacles are long with sharp ends. *Size:* Column diameter to about 2.5 inches (60 mm). *Range and Habitat:* Bodega Bay, California to Baja California on hard substrate low Intertidal to about 60 ft (18 m).

62. MOONGLOW ANEMONE

Anthopleura artemisia

Identification: The abundant, long tapering tentacles are white, pale pink, red, orange, or dark blue, either in solid color or in patterns. The upper column of this solitary anemone has well developed tubercles. *Size:* Column diameter to about 2 inches (50 mm). Tentacular crown diameter to almost 3 inches (75 mm). *Range and Habitat:* Alaska to southern California, attached to rocks that are buried in the sand in bays. Intertidal to about 100 ft (30 m).

63. SWIMMING ANEMONE

Stomphia coccinea

Identification: This anemone has a column that is as wide as its height. It is usually orange, with a white spot at the base of each tentacle. The tentacles have white bands. Size: Diameter to about 5 inches (125 mm). *Range and Habitat:* Circumpolar on this coast from Alaska to southern California. In very deep water on rocks; shallow subtidal in Washington and Alaska waters.

64. ANEMONE

Metridium senile

Identification: This small, aggregating anemone lacks a lobed oral disc. The tentacles are solitary, short slender and tapered. Colors range from white to orange, cream, brown or tan. *Size:* Height to about 4 inches (100 mm). Column diameter to about 2 inches (50 mm). *Range and Habitat:* Circumpolar; on our coast, from Alaska to southern California. On rocks wharf pilings and other man-made structures. From intertidal to depths of less than 100 ft (30 m).

65. WHITE-PLUMED ANEMONE

Metridium farcimen

Identification: This large, white, solitary anemone has a lobed oral disc. The tentacles are small and appear fluffy. *Size:* Height to about 20 inches (50 cm). Column diameter to about 4 inches (100 mm). *Range and Habitat:* Alaska to Santa Catalina Island, California; on reefs, wrecks, and other structures. Subtidal depths to at least 670 ft (200 m). *Natural History:* This anemone only reproduce sexually.

66. CLUB-TIPPED ANEMONE

Corynactis californica

Identification: These distinctive, aggregating anemones have tentacles with bulbous tips. Color varies from orange, red, purple, pink, or brown to almost white. *Size:* Diameter of tentacular crown to about 1 inch (25 mm) Height to about 3 inches (75 mm). *Range and Habitat:* Nootka Sound, British Columbia to San Martin Island, Baja California. On rocks, particularly where currents prevail; low intertidal to about 150 ft (45 m). *Natural History:* This anemone reproduce asexually; the clones are products of longitudinal fission of adults.

67. SAND ANEMONE

***Phyllactis* sp.**

Identification: This sand dwelling anemone has short, slender, clear, tapering tentacles that have black stripes. The oral disc usually is level with the sand surface and is covered with sand particles. *Size:* Tentacular crown diameter to about 3 inches (75 mm). *Range and Habitat:* Southern California, on subtidal sand and bedrock bottoms.

68. GHOST ANEMONE

Diadumene leucolena

Identification: These small anemones have long, pale-green tentacles. *Size*: Diameter to about 1 inch (25 cm) and height to about 2 inches (5 cm). *Range and Habitat*: This species was introduced from the North Atlantic. On this coast, located from Oregon to Baja California in bays and estuaries.

69. YELLOW ZOANTHID

Parazoanthus lucificum

Identification: These golden yellow zoanthids are only found on gorgonians. *Size:* Width of column to about 0.25 inches (6 mm). Height to about 0.5 inches (12 mm). *Range and Habitat:* Santa Catalina Island, California to San Benitos Islands, Baja California; on gorgonians. Subtidal depths to about 80 ft (24 m).

70. ZOANTHID

(undescribed species)

Identification: Very similar to # 69, but the aggregations are attached to rocks, not on gorgonians. *Size:* Width of column about 0.25 inches (10 mm). Height to about 0.5 inches (15 mm). *Range and Habitat:* Santa Catalina Island, California.

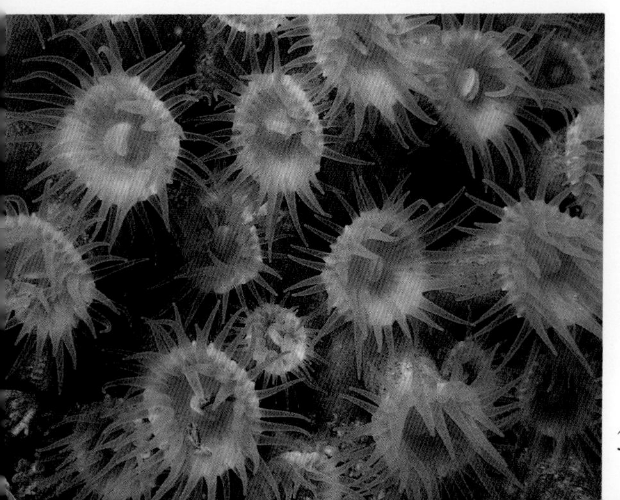

71. YELLOW ZOANTHID

Epizoanthus scotinus

Identification: The color of this zoanthid ranges from tan to brown or light yellow. The tentacles are light yellow or white. *Size:* Diameter to about 0.75 (20 mm). *Range and Habitat:* Southern Alaska to southern California (northern Channel Islands); on rocks from low intertidal to depths of at least 180 ft (54 m). *Natural History:* Reproduce sexually and asexually producing dense aggregations of clones.

72. SEA PANSY

Renilla koellikeri

Identification: The heart shaped, purple disc consists of a mass of attached transparent polyps. It has a fleshy peduncle that anchors the animal in the sand. The polyps have eight tentacles. *Size:* Diameter of disc to about 5 inches (125 mm). *Range and Habitat:* Southern California to Cedros Island, Baja California. On sand, in shallow waters. *Natural History:* The nudibranch *Armina californica,* as well as the sand star *Astropecten armatus,* feed on the polyps.The colony luminates when distrubed.

73. WHITE SEA PEN

Stylatula elongata

Identification: A slender, rough to touch, sea pen with a whitish-gray axis and fluffy plume-like gray or greenish lateral branches. Each of the tiny feeding polyps has eight feather-like tentacles. *Size:* Height to about 24 inches (60 cm). *Range and Habitat:* British Columbia to Isla Cedros, Baja California, on sandy or mud bottoms to depths of 200 ft (60 m) or more. May occur in the Gulf of California. *Natural History:* The polyps feed on drifting plankton. This sea pen luminesces when disturbed.

74. ORANGE SEA PEN

Ptilosarcus gurneyi

Identification:. A large, orange-to-yellow sea pen with a fleshy stalk and feather-like branches. The buried portion consists of a bulbous-shaped peduncle. Size: Height to about 18 inches (46 cm). *Range and Habitat:* Alaska to southern California, on sand bottoms. Depth range from about 25 ft (8 m) off Alaska and British Columbia, to at least 300 ft (100 m) off southern California. *Natural History:* This sea pen produces a greenish lumi-nescence when disturbed. Several species of nudibranch prey on this sea pen.

75. CUP CORAL

Astrangia lajollaensis

Identification: These small cup corals form large colonies. The stony cups are brownish-orange and the short tentacles are yellowish-orange. *Size:* Cups to about 0.5 inches (12 mm) in diameter. *Range and Habitat:* Monterey Bay, California to Baja California. Aggregations occur on lower sides of rocks and reefs in depths of 40 ft (12 m) or deeper.

76. ORANGE CUP CORAL

Balanophyllia elegans

Identification: Each solitary, orange, cup-shaped coral has orange tentacles. Each tentacle bears wart-like nodules containing batteries of stinging cells (nematocysts). *Size:* Diameter of cup about 1 inch (25 mm). *Range and Habitat:* British Columbia to central Baja California. Attached to rocks in depths from the low intertidal to at least 160 ft (48 m). *Natural History:* Sexes are separate. The eggs are fertilized and develop within the female's gastrovascular cavity. The worm-like planula larvae are released during spring and summer.

77. BROWN CUP CORAL

Paracyathus stearnsi

Identification: The brown, solitary cups have long, almost clear, tentacles. *Size*: Cups may reach 1.5 inches (38 mm) in diameter. *Range and Habitat:* British Columbia to Cedros Island, Baja California. On lower sides of rocky reefs, in depths of 30 to 200 ft (9 to 60 m).

78. COLONIAL CUP CORAL

Coenocyathus bowersi

Identification: These cup corals occur in colonies; each orange to pinkish cluster contains elongated cup corals with short tentacles. *Size:* Width of colony to about 6 inches (150 mm). Height to about 3 inches (75 mm). *Range and Habitat:* Monterey Bay, California to Baja California and the Sea of Cortez. In caves and on sides of rocks in depths of 30 to 490 ft (9 to 146 m).

79. TUBE-DWELLING ANEMONE

Pachycerianthus fimbriatus

Identification: This bottom dwelling anemone has long, slender, outer tentacles and shorter, inner tentacles. The soft body is protected by a secreted parchment-like tube. Color of the tentacles may be creamy white, brown, black, or orange. *Size:* Height to about 12 inches (300 mm). Diameter to about 2.5 inches (38 mm). *Range and Habitat:* Alaska to Isla San Martin, Baja California; on sand and soft mud bottoms. Low intertidal to depths of at least 70 ft (21 m). *Natural History:* The nudibranch *Dendronotus iris* feeds on this anemone; however, feeding attacks seldom kill the anemone.

80. OCTOCORAL

Alcyonium rudyi

Identification: Each polyp of these cream to pink soft coral colonies has eight long, branched tentacles. *Size:* Width of colony to at least 4 inches (100 mm). Polyp height about 1 inch (25 mm). *Range and Habitat:* Cape Flattery, Washington to southern California; on undersurface of rocks, occasionally on decorator crabs. Very low intertidal and shallow subtidal.

81. SEA STRAWBERRY

Gersemia rubiformis

Identification: The thick, red, soft lobes of this soft coral have white polyps. *Size:* Height of colony to 4 inches (100 mm). Base underneath to about 12 inches (300 mm). *Range and Habitat:* Temperate Atlantic and Pacific oceans. On this coast from Bering Sea to Point Arena, California. Intertidal to about 50 ft (15 m) on offshore pinnacles or areas of strong currents. *Natural History:* The nudibranch *Tochuina tetraquetra* feeds on this soft coral.

82. PURPLE GORGONIAN

Eugorgia rubens

Identification: This gorgonian sea fan has slender purple or violet branches; the polyps are white. The branches form an interwoven pattern on one plane. *Size:* Height to at least 12 inches (30 cm). *Range and Habitat:* Southern California to Baja California. Attached to rocks, in depths of 80 to 100 ft (24 to 30 m). It is common around the San Benito Islands off Baja California.

83. ORANGE GORGONIAN

Adelogorgia phyllosclera

Identification: A sea fan with slender orange branches, on a single plane and with yellow polyps. *Size:* Colony may reach 24 inches (60 cm) in height. *Range and Habitat:* Southern California to Baja California. On rocks, in depths from about 100 to 1000 ft (30 to 300 m).

84. RED GORGONIAN

Lophogorgia chilensis

Identification: This distinctive sea fan has red branches with white polyps. The branches are not in a single plane. *Size:* Height to about 36 inches (90 cm). *Range and Habitat:* Monterey Bay to Isla Cedros, Baja California. In depths of about 50 to 200 ft (15 to 60 m). *Natural History:* The ovulid snail *Delonovolva* (# 125) lives and feeds on the branches of this gorgonian. The red gorgonian is often colonized by the zoanthid, *Parazoanthus lucificum,* with the ultimate result of the death of all or most of the red gorgonian polyps.

85. BROWN GORGONIAN

Muricea fruticosa

Identification: The thick brown branches with white polyps usually branch in one plane. Size: Height to about 3 ft (90 cm). *Range and Habitat:* Point Conception, California to Cedros Island, Baja California. On reefs, pipelines, wharfs, and wrecks in depths of 50 to 100 ft (15 to 30 m). *Natural History:* This is one of the most common gorgonians in southern California. The colonies are able to survive in some of the most polluted nearshore waters.

86. CALIFORNIA GOLDEN GORGONIAN

Muricea californica

Identification: The thick brown branches have yellow polyps. The yellow polyps distinguish the gorgonian from the brown gorgonian. *Size:* Height to about 3 ft (90 cm). *Range and Habitat:* Point Conception, California to outer Baja California and Sea of Cortez. On rocks and other hard substrate; depths range from about 40 to 100 ft (12 to 30 m). *Natural History:* As the colony grows, annual growth rings are formed in the skeleton. Each colony is a separate sex.

87. GORGONIAN

Calcigorgia spiculifera

Identification: This northern gorgonian is similar to both species of *Muricea* (# 85 and 86), but it only occurs north of Puget Sound, Washington. Size: Height to about 10 inches (250 mm). *Range and Habitat:* Northeastern Alaska to British Columbia. On rocks, in subtidal depths below about 80 ft (24 m).

PHYLUM CTENOPHORA

Comb Jellies

All of the members of this phylum are free-swimming members of the marine zooplankton community. Their main source of locomotion is the ocean currents within which they are suspended; however, they can move about on a limited scale using their eight rows of cilia (comb rows). These transparent animals are often confused with jellyfish, but they differ in two ways: they lack the nematocysts (stinging cells) of jellyfish and they do not have tentacles around the mouth. The body consists of a mouth at one end and an anus at the other. The simple digestive system consists of a pharnyx and stomach. The reproductive system contains both ovaries and testes. The egg and sperm, when fertilization occurs, develop into a larval form that resembles the adult. Some species have tentacles that are used to capture food. All of the comb jellies are predators on other zooplankton. They, in turn, are fed on by some species of fishes and sea turtles. Twenty species of comb jellies have been identified from the waters of Alaska, British Columbia, Washington, Oregon, and California.

CLASS NUDA

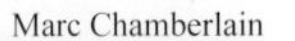

Marc Chamberlain

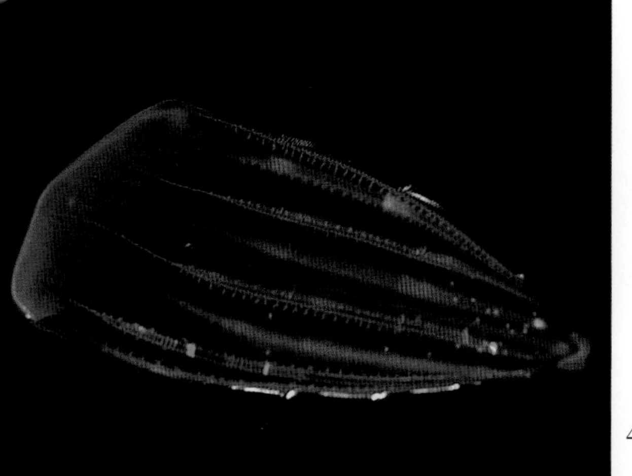

88. COMB JELLY

***Beroe* sp.**

Identification: The elongated, saclike body is pointed at one end. *Size:* Lengths to about 6 inches (150 mm). *Range and Habitat:* Alaska to California. Planktonic; found near the surface.

PHYLUM PLATYHELMINTHES

Flatworms

There are three classes of flatworms but only one, Class Turbellaria, is free-living, i.e. not parasitic. Of the two groups in this class, only the polyclads are large enough to be readily observed by divers and tidepoolers. Polyclads are primarily marine worms, although there are a few freshwater forms.

The general anatomy of the flatworms includes a branched digestive system, a mouth located near the middle, on the underside of the body, a set of male and female sex organs, and primitive eyes, usually on short stalks. Although these worms are hermaphroditic, they cannot undergo self-fertilization. The internally fertilized eggs are deposited in gelatinous strands that attach to the substrate. The eggs either develop directly into a juvenile worm or undergo a free-swimming larval phase. All these worms are carnivores, feeding on small crustaceans and bryozoans. Approximately 42 species of polyclads have been identified from this coastal area.

89. POLYCLAD FLATWORM

Pseudoceros luteus

Identification: A whitish flatworm with a light, white, ruffled margin. *Size:* Length to 4 inches (10 cm). *Range and Habitat:* British Columbia to southern California. On and under rocks; from low intertidal to shallow subtidal.

90. POLYCLAD FLATWORM

Pseudoceros montereyensis

Identification: White with numerous black spots and a black center stripe. Tentacles are black banded; margin slightly ruffled. *Size:* Length to about 3.5 inches (90 mm). *Range and Habitat:* Monterey Bay to southern California. Under stones; on intertidal and subtidal rocky coast.

Marc Chamberlain

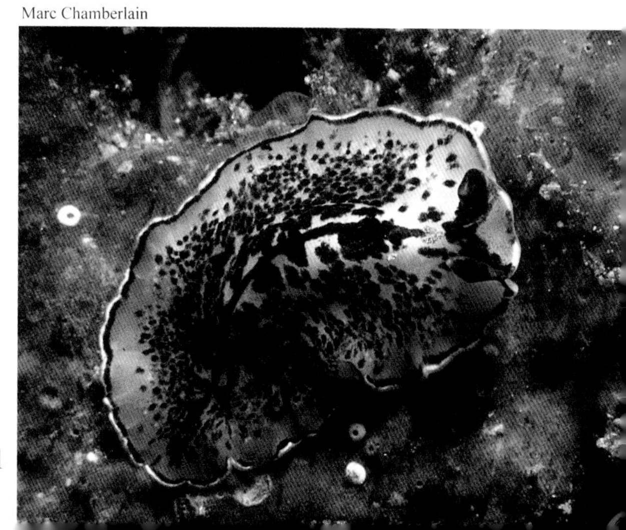

Marc Chamberlain

91. POLYCLAD FLATWORM

Prostheceraeus bellostriatus

Identification: This distinctive flatworm has an orange central band with parallel black stripes running from the anterior end to the posterior end. The border is also orange and the tentacles are black. *Size:* Length to about 1.5 inches (35 mm). Width to about 1 inch (25 mm). *Range and Habitat:* Monterey Bay to southern California. Under rocks and on wharf pilings; intertidal and shallow subtidal.

Marc Chamberlain

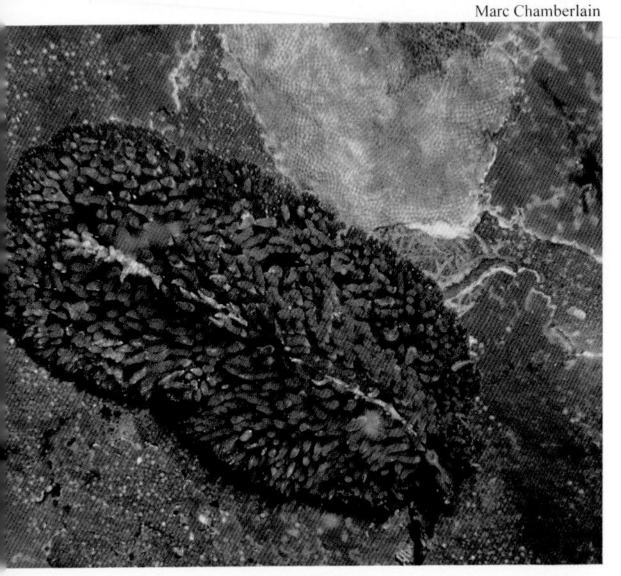

92. POLYCLAD FLATWORM

Thysanozoon sandiegense

Identification: Dorsal surface black with lighter mottling on papillae. *Size:* Length to about 1.25 inches (30 mm). *Range and Habitat:* Southern California. Under rocks; found in low intertidal and shallow subtidal.

93. POLYCLAD FLATWORM

Eurylepta californica

Identification: Gray-white with black lines radiating from white center line. *Size*: Length to about 1.25 inches (30 mm). *Range and Habitat:* Nootuz Sound, British Columbia to southern California. Under rocks; intertidal and shallow subtidal.

PHYLUM ANNELIDA
Worms

This is one of the largest phyla of marine invertebrates in the world. They inhabit all of the world's oceans and seas. Although nearly all of the polychaetes are marine, a few are found in brackish environments and in freshwater, as well as on land in damp soil.

Polychaete worms are characterized by having a segmented body, which contains a head at one end and an anus at the opposite end. Each segment of the elongated body is called a somite. The somites are separated by grooves. All the species covered in the guide have tentacles or gills associated with the mouth, and a small brain located near the gut. Some species have visible eyes. There are blood vessels that form a primitive circulatory system. All species have bristles (setae) on at least some of the segments. Polychaetes can be found in all types of marine habitats including the pelagic region. Although polychaetes feed on all types of animals and plants, those with tentacles are usually filter feeders, using their tentacles to collect small animals and detritus from the water column. Some of these worms can regenerate either the head or tail end or both.

The sexes are separate; sperm and eggs are released into the water for fertilization. The fertilized eggs develop into trochophore larvae. Growth in the young worms occurs with the addition of body segments. There are approximately 830 species of polychaetes on this coast. Only a few of the most common species are included here.

CLASS POLYCHAETA
Polychaete Worms

94. CHAETOPTERID WORM
Phyllochaetopterus prolifica

Identification: The large, bush-like cluster of these tube dwelling worms is distinctive. *Size*: Diameter of tubes is about 1/16 inch (1 mm). Clusters may reach 12 inches (30 cm) in height. *Range and Habitat*: British Columbia to southern California, shallow subtidal to at least 80 ft. (24 m). On rocks, pier pilings and floating docks. *Natural History*: These worms reproduce asexually by dividing, as well as secually; eggs and sperm are released into the water.

Alan Studley

95. COLONIAL TUBE WORMS

Dodecaceria fewkesi

Identification: Colonies consist of clusters of calcareous tubes. The exposed tentacles are dark brown to black and are composed of a pair of polyps and 11 pairs of branchial filaments. *Size:* Cluster to about 3 ft (1 m) in diameter. Diameter of the tube to about 0.25 inch (5 mm). Length of worm to about 1.5 inches (38 mm). *Range and Habitat:* British Columbia to southern California. On rocks and dock pilings, intertidal to shallow subtidal.

96. COLONIAL SAND TUBE WORM

Phragmatopoma californica

Identification: The extensive colonies consist of tubes constructed of fine sand that the worm cements together. The short tentacles are lavender. These worms possess an operculum to close off the tube. *Size:* Diameter of tubes to about 0.5 inch (12 mm). Colonies to at least 6 ft (2 m) across. *Range and Habitat:* Central California to northern Baja California. Honeycombed colonies formed on and against rocks, particularly in areas of high sand transport. Intertidal depths to about 240 ft (72 m).

97. ORNATE TUBE WORM

Diopatra ornata

Identification: The parchment tubes of these worms are usually completely covered with attached pieces of shells, algae, sticks, and other debris. *Size:* Diameter of tube to about 0.75 inch (18 mm). *Range and Habitat:* British Columbia to Baja California; in sand and rubble subtidal bottoms, usually near rocks, to about 300 ft (90 m).

98. FIBER TUBE WORM

Pista elongata

Identification: The solitary tubes of this worm terminate in a distinctive globular shaped, reticulated net-work of fibers. *Size:* Height of tube to about 3 inches (75 mm). *Range and Habitat:* British Columbia to San Diego, California; attached to rocks, from middle intertidal to subtidal depths of about 40 ft (12 m).

99. NORTHERN FEATHER DUSTER WORM

Eudistylia vancouveri

Identification: The distinctive tentacular crown is dark maroon with dark green stripes. The large parchment tubes are light grey. *Size:* Tentacular crown to about 2 inches (50 mm) in diameter. Tube length to about 18 inches (45 cm). *Range and Habitat:* Alaska to northern California. On rocks and dock pilings, at shallow subtidal depths.

100. FEATHER DUSTER WORM

Eudistylia polymorpha

Identification: Tentacular crown consists of a plume of branched gills. Color varies from light tan to orange. *Size:* Tentacular crown diameter to about 3 inches (75 cm). Height of tubes to about 11 inches (275 mm). *Range and Habitat:* Alaska to southern California, attached to rocks and dock pilings from low intertidal to subtidal depths of about 130 ft (39 m). *Natural History:* The tentacles are used to obtain oxygen as well as for filter feeding.

101. CHRISTMAS TREE WORM

Spirobranchus spinosus

Identification: Branchial plume consists of three concentric whorls in a variety of colors. *Size:* Width of plume to about 0.75 inches (18 mm). Height of plume to about 1 inch (25 mm). *Range and Habitat:* Central California to Baja California. Tubes attached to rocks from intertidal to depths of about 40 ft (12 m).

102. FRAGILE TUBE WORM

Salmacina tribranchiata

Identification: This worm constructs small whitish tubes that form tangled complex masses. The worms do not have an operculum. *Size:* Tube diameter less than 0.12 inch (2 mm). Masses to about 8 inches (200 mm) across. *Range and Habitat:* British Columbia to southern California. Tubes attached to sides and undersides of rocks; from lower intertidal to depths of about 50 ft (15 m). *Natural History:* The fertilized eggs develop within the tube. These worms also reproduce asexually by budding and transverse fission.

103. SERPULID WORM

Serpula vermicularis

Identification: These worms construct white calcareous tubes that are often coiled. The white crown of gills (branchial crown) consists of 40 pairs of plumes. This worm has a funnel shaped operculum. *Size:* Tube length to about 4 inches (100 mm). Width of branchial crown to about 1 inch (25 mm). *Range and Habitat:* Northwestern Pacific to Alaska and south to San Diego, California, also occurs in Atlantic and Indian oceans. Tubes are attached to sides and top of rocks. In the low intertidal zone to depths of at least 300 ft (100 m).

104. SABELLID WORM

Myxicola infundibulum

Identification: The whorl of gill plumes are connected by a membrane, which forms a sort of funnel. The color of the plumes ranges from yellow to green-purple or brown. These plume worms inhabit a secreted gelatinous tube. *Size:* Worm length to about 3.5 inches (90 mm). Width of plume to about 1 inch (25 mm). *Range and Habitat:* Bering Sea to California. Wedged in fouling organisms on piers and rocks and embedded in soft bottom sediments; subtidal depths to 600 ft (180 m).

105. TEREBELLID WORM

Thelepus crispus

Identification: The tube is a thin membrane covered with sand. The feeding tentacles are very long, transparent, and at least 12 inches (300 mm) in length. *Size:* Worm body length to about 11 inches (275 mm). *Range and Habitat:* Alaska to southern California. In stable sand and mud bottoms and under rocks; from middle and low intertidal to depths of about 50 ft (15 m).

PHYLUM MOLLUSCA

Snails, Sea Slugs, Chitons, Clams, Squid, and Octopuses

This very large group of animals is found in all of the world's marine, aquatic, and terrestrial habitats. Their fossil record dates back over 500 million years. The vast majority of these shelled, soft-bodied invertebrates are mobile. Mollusc anatomy varies greatly. Generally the body can be divided into a head, foot, and visceral mass or hump. However, bivalve molluscs lack a distinctive head. Other groups lack the foot. Most have a shell, at least during some period of their life, but a few never produce a shell. All of the organ systems present in more advanced animals are present in molluscs including: a complex nervous system, a circulatory system, and digestive and excretory systems. Many species have primitive eyes. Some, such as the squid and octopuses have very complex nervous systems, including a brain.

This group contains all types of feeding strategies, from filter feeders to carnivores. All of the molluscs reproduce sexually. Individuals of some species have both male and female sex organs (hermaphrodites).

Molluscs provide food for a variety of other invertebrates, fishes, birds, and mammals. Of the seven classes of molluscs, four are covered in. this field guide: Polyplacophora, Gastropoda, Bivalvia, and Cephalopoda. These four classes are represented by at least 77, 918, 391, and 60 species, respectively, in the area covered by this guide.

CLASS POLYPLACOPHORA
Chitons

106. LINED CHITON
Tonicella lineata

Identification: Easily distinguished from other chitons by the zigzag lines and bands of various colors, including reds and blues. *Size:* Length to about 2 inches (50 mm). *Range and Habitat:* Northern Japan and Aleutian Islands to San Diego, California; on rocks, usually with encrusting coralline algae. From low intertidal to depths of about 100 ft (30 m). *Natural History:* Often found under purple sea urchins (# 226). They are preyed on by the sea stars *Pisaster ochraceus* and *Leptasterias hexactis.*

107. GUMBOOT CHITON
Cryptochiton stelleri

Identification: This is the largest chiton in the world and is distinguished by the leathery covering (mantle) covering the valves. They are usually brick-red or reddish-brown. *Size:* Length to about 13 inches (325 mm). *Range and Habitat:* Northern Japan and Aleutian Islands to San Nicholas Island, California, on rocky as well as soft bottoms. Low intertidal to about 60 ft (18 m). *Natural History:* These chitons grow slowly and may live 20 years or more. They feed on red algae.

CLASS GASTROPODA
Snails and Slugs

108. PINK ABALONE
Haliotis corrugata
Identification: The shell is almost round, with a corrugated, scalloped edge. Two to four raised shell holes remain open. *Size:* Length to about 10 inches (250 mm). Legal size is 6 inches (150 mm). *Range and Habitat:* Point Conception, California to Bahia Tortugas, Baja California, in rocky crevices and on sides of rocks. Low intertidal to about 180 ft (54 m). *Natural History:* The most common abalone in southern California.

109. BLACK ABALONE
Haliotis cracherodii
Identification: The smooth outer shell's color ranges from dark blue to black. Five to seven of the shell holes are usually open. Size: Length to about 8 inches (200 mm). Legal size in California is 5 inches (125 mm). *Range and Habitat:* Mendocino, California to Cabo San Lucas, Baja California. On rocks and in crevices, from intertidal zone to about 20 ft (6 m).

110. GREEN ABALONE
Haliotis fulgens
Idenitifcation: The exterior shell is olive green to reddish. Five to seven circular shell holes are open. The inside of the shell is an opalescent dark green, blue, and lavender. *Size:* Length to about 10 inches (250 mm). Legal size is 6 inches (150 mm). *Range and Habitat:* Point Conception, California to Bahia Magdalena, Baja California. In crevices and on rocks; low intertidal to about 60 ft (18m).

111. PINTO ABALONE

Haliotis kamtschatkana

Identification: The outer shell color is reddish with white and blue markings. Three to six shell holes are open. The shell-like openings are slightly raised. *Size:* Length to about 6 inches (150 mm). Legal size in California is 11.5 inches (340 mm). *Range and Habitat:* Sitka, Alaska to Point Conception, California. On rocky substrate; low intertidal to about 115ft (34 m).

112. RED ABALONE

Haliotis rufescens

Identification: The outer shell is usually brick-red, occasionally with bands of green or white. Three or four of the holes in the shell remain open. *Size:* Length to 12.3 inches (315 mm). Legal size in California for sport harvest is 7 inches (175 mm). *Range and Habitat:* Sunset Bay, Oregon to Bahia Tortugas, Baja California. On rocks; occasionally will move across sand or gravel bottoms. From low intertidal to about 80 ft (24 m). *Natural History:* This is the most sought after of all the abalone along the Pacific Coast. Red abalone are slow growing; it may take 12 to 14 years to reach legal size.

Ron McPeak

113. WHITE ABALONE

Haliotis sorenseni

Identification: Raised shell holes number from three to five. The edge of the foot (epipodium) is a light mottled yellow-green and beige. Size: Length to about 10 inches (250 mm). Legal size is 6 inches (150 mm). *Range and Habitat:* Point Conception, California, to central Baja California. On rocks; in subtidal depths of 70 to about 200 ft (21 to 60m).

114. FLAT ABALONE

Haliotis walallensis

Identification: The shell is more flat and oblong than any of the other species of abalone on this coast. The number of open holes in the shell ranges from four to eight. *Size:* Length to about 7 inches (175 mm). Legal size in California is 4 inches (100 mm). *Range and Habitat:* British Columbia to La Jolla, California. On rocks or in crevices, low intertidal to about 70 ft (21m). *Natural History:* The sea otter and cabezon *Scorpaenichthys marmoratus,* are major predators.

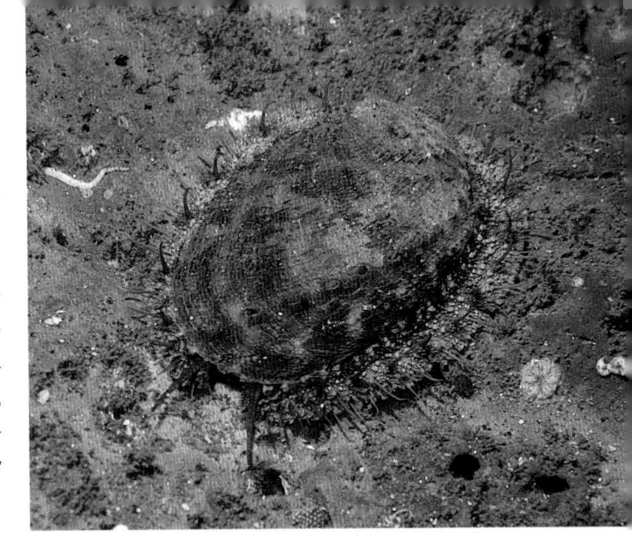

115. ROUGH KEYHOLE LIMPET

Diodora aspera

Identification: The mantle covers much of the shell. Mantle color varies from white to red. The conical shell of this limpet is exposed and has a circular opening in the top. *Size:* Length is about 3 inches (75 mm). *Range and Habitat:* Afognak Island, Alaska, to Camalu, Baja California, on rocky substrate. Low intertidal to about 50 ft (15 m). *Natural History:* These limpets use their mantles to foil attacks by sea stars.

116. GIANT KEYHOLE LIMPET

Megathura crenulata

Identification: The mantle usually covers most or all of the shell. The shell has a large oval opening in the center. Mantle color varies from black to mottled grey. *Size:* Length of shell to about 5 inches (125 mm). *Range and Habitat:* Mendocino County, California to Isla Asuncion, Baja California, on rocky substrate. Low intertidal to about 110 ft (33 m). *Natural History:* These large limpets feed on algae and tunicates.

117. WHITECAP LIMPET

Acmaea mitra

Identification: The conical, white shell is often covered with encrusting red coralline algae. *Size:* Length to about 1.5 inches (35 mm). Height to about 1 inch (30 mm). *Range and Habitat:* Aleutian Islands, Alaska to Isla San Martin, Baja California. On rocks; low intertidal to about 100 ft (30 m).

118. BLUE-RING TOP SNAIL

Calliostoma annulatum

Identification: The distinctive gold and blue colored shell of this snail is sometimes covered with other organisms. The foot is a bright orange-yellow with brown blotches. *Size:* Shell diameter to about 1.2 inches (30 mm). *Range and Habitat:* Forrester Island, Alaska to Isla San Geronimo, Baja California. On rocks and kelp; low intertidal zone to about 140 ft (42 m). *Natural History:* These colorful snails feed on hydroids, bryozoans, diatoms, detritus, and copepods, as well as kelp.

119. BLUE TOP SNAIL

Calliostoma ligatum

Identification: This conical shell is about as high as it is wide. It has light tan spiral ridges on a brown background. The brown foot is orange on the bottom. *Size:* Shell diameter to about 1 inch (25 mm). *Range and Habitat:* Prince William Sound, Alaska to San Diego, California. On kelp and rocks, low intertidal to about 60 ft (18m). *Natural History:* This kelp-eating snail also feeds on diatoms, detrius, and even hydroids.

120. RED TURBAN SNAIL

Lithopoma gibberosum

(Formerly ***Astraea gibberosa)***

Identification: The low, spiral shell is usually red-brown. The oval-shaped, calcareus operculum has a shallow groove and is smooth. *Size:* Shell diameter to about 2 inches (50 mm). *Range and Habitat:* Queen Charlotte Islands, British Columbia to Bahia Magdalena, Baja California. On rocks; low intertidal to about 270 ft (80 m).

121. WAVY TURBAN SNAIL

Lithopoma undosum

(Formerly ***Astraea undosa)***

Identification: The tan-colored shell is heavily sculptured and the base of the shell has spiral cords. The ornate operculum has large, rough ridges. *Size:* Shell diameter to about 4.5 inches (110 mm). *Range and Habitat:* Point Conception, California to Isla Asunción, Baja California. On rocks, usually in kelp beds; low intertidal to about 70 ft (21 m).

122. NORRIS'S TOP SNAIL

Norrisia norrisi

Identification: The flattened spiral shell is similar to *Polinices* (# 124), but is a dark red-brown. The foot is a bright red. *Size:* Shell diameter to about 2.5 inches (55 mm). *Range and Habitat:* Point Conception, California, to Isla Asunción, Baja California. On rocks and kelp, in depths from 10 to about 100 ft (3 to 36 m). *Natural History:* Usually associated with giant kelp, *Macrocystis,* on which it also feeds.

123. SCALED WORM SHELL

Serpulorbis squamigerus

Identifcation: The shell consists of a partially coiled tube that is attached to the substrate. There is no operculum. *Size*: Length of tube to about 5 inches (125 mm). Tube diameter to about 0.5 inches (12 mm). *Range and Habitat:* Gualala, Sonoma County, California, to Baja California. Attached to rocks; intertidal to about 100 ft (30 m). *Natural History:* These tube-shelled molluscs often live in aggregations of 600 or more individuals. They are filter feeders, using mucous to collect drifting plankton.

124. LEWIS' MOON SNAIL

Polinices lewisii

Identification: The smooth, brownish colored shell is globular shaped. The foot is gray-brown and is capable of almost covering the shell. *Size:* Shell diameter to about 5 inches (125 mm). *Range and Habitat:* Vancouver Island, British Columbia, to Isla San Geronimo, Baja California. On soft bottoms, shallow bays, and open ocean, to depths of about 500 ft (150 m). Feeds on clams.

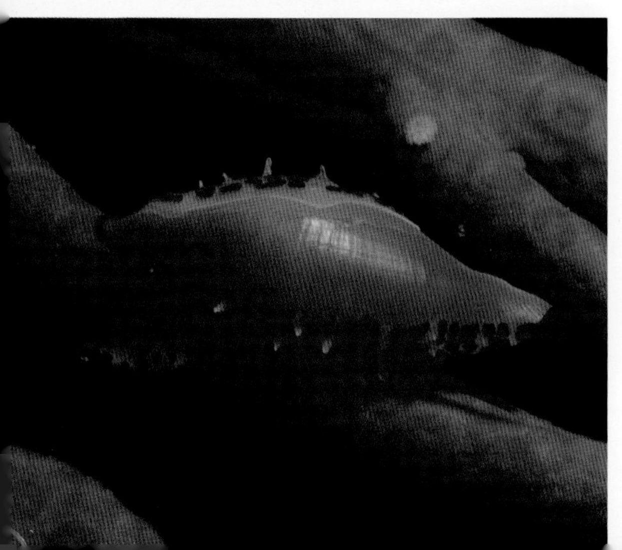

125. SIMNIA

Delonovolva aequalis

Identification: The smooth rose-colored shell has an opening that extends from tip to tip as in cowries. It's papillated, colorful mantle sometimes covers the shell entirely. *Size:* Shell length to about 1 inch (25 mm). *Range and Habitat:* Monterey Bay, California to Isla San Martin, Baja California. On red gorgonians (# 84), in depths of 30 to at least 100 ft (9 to 30 m).

126. CHESTNUT COWRY

Cypraea spadicea

Identification: A smooth shell with a brown dorsal surface and gray or white margins and ventral side. This is the only cowry that occurs in this area. The foot and mantle are orange-brown with dark spots. *Size:* Length of shell to about 3 inches (75 mm). *Range and Habitat:* Monterey, California to Isla Cedros, Baja California. On rocks, usually in kelp beds; low intertidal to about 150 ft (45 m). *Natural History:* This cowry is a scavenger and carnivore; it feeds on anemones, sponges, tunicates, and snail eggs.

127. HAIRY TRITON

Fusitriton oregonensis

Identifcation: The large brown shell is usually covered with a hairy periosticum. The shell has about six whorls. *Size:* Height of shell can reach 6 inches (150 mm). *Range and Habitat:* Aleutian Islands to southern California. On rocky as well as soft bottoms; in subtidal depths to about 600 ft (180 m). *Natural History:* These carnivorous snails feed on sea urchins or by scavenging.

128. KELLET'S WHELK

Kelletia kelletii

Identification: The robust shell of this snail has heavy sculpturing crossed by thin spiral lines. Shell color is white or gray. Animal skin is yellow, mottled with black and white. *Size:* Length of shell to about 7 inches (175 mm). *Range and Habitat:* Monterey Bay, California, to Isla Asuncion, Baja California. In kelp beds on rocky and soft bottoms; low intertidal to about 230 ft (69 m).

129. LEAFY HORNMOUTH

Ceratostoma foliatum

Identification: This snail has three wing-like processes that protrude from the central shell, one on top and one on each side. There is also a tooth on the outside edge of the opening. *Size:* Length of shell to about 3.5 inches (90 mm). *Range and Habitat:* Alaska to San Diego, California. On rocky substrate; low intertidal to about 150 ft (45 m).

130. IDA'S MITER

Mitra idae

Identifcation: The dark brown, relatively smooth shell is covered with a black periostracum. The foot is white. *Size:* Length of shell to about 3 inches (75 mm). *Range and Habitat:* Crescent City, California to Isla Cedros, Baja California. On rocks; low intertidal to about 70 ft (21 m).

131. CALIFORNIA CONE

Conus californicus

Identification: The only species of cone that occurs in the area covered by this field guide. The light brown shell is smooth. *Size:* Shell length to about 1.5 inches (40 mm). *Range and Habitat:* Farallon Islands, California to Bahia Magdalena, Baja California. On sandy as well as rocky bottoms; low intertidal to about 150 ft (45 m). *Natural History:* Like all cones, this species subdues its prey by "harpooning" it with a poison laden, spear-like tooth.

CLASS OPISTHOBRANCHIA
Opisthobranchs (Sea Slugs)

132. NAVANAX
Navanax inermis
Identification: *Navanax* has an internal shell that is not visible. The longitudinal yellow stripes and yellow-and-blue spots on the body are distinctive. *Size*: Length to about 9 inches (22.5 cm). *Range and Habitat*: Bolinas Lagoon, California to Laguna Manuela, Mexico; including to Sea of Cortez. On sand, mud and rocky bottoms. *Natural History*: This predator feeds on other opisthobranchs.

133. CALIFORNIA SEA HARE
Aplysia californica
Identification: *Aplysia* has dorsal flap-like extensions (parapodia) of the foot. The large rhinophores resemble rabbit ears, hence the common name. Color varies from light to medium brown, mottled with red areas and reticulating patterns. *Size*: Length to about 18 inches (45 cm). *Range and Habitat*: Hachiyo Island, Japan to Baja California and Sea of Cortez and El Salvador. Low intertidal to depths of about 80 ft. (22 m). *Natural History*: This species feeds on various species of algae, including *Gracillaria*, *Ulva* and *Enteromorpha*.

134. CATALINA TRIOPHA
Triopha catalinae
Identification: There is a frontal veil on this distinctive nudibranch. The veil has 8 to 12 orange projections. The white body has numerous orange tubercles; the rhinopores are orange to red. *Size*: Length to about 3 inches (70 mm). *Range and Habitat*: Amchitka Island, Alaska to El Tomutal, Baja California. Low intertidal to about 60 ft. (17 m). *Natural History*: This nudibranch feeds on bryozoans.

135. MONTEREY DORID

Doris montereyensis

(Previously *Archidoris montereyensis*)

Identification: This yellow nudibranch differs from #136 by having some of the tubercles covered by black spots. The color pattern varies. *Size*: Length to 6 inches (15 cm). *Range and Habitat*: Point Craven, Alaska to Sea of Cortez. Also reported from Chile and Argentina. On hard and soft bottoms to depths of less than 100 ft. (28 m).

136. SEA LEMON

Peltodoris nobilis

(previously ***Anisodoris nobilis***)

Identification: The black spots on this yellow nudibranch do not extend on to the tubercles, and the branchial plume is white. *Size*: Length to about 8 inches (20 cm). *Range and Habitat*: Kodiak Island, Alaska to Coronado Islands, Baja California. On hard and soft bottoms to depths of 750 ft. (208 m). *Natural History*: This nudibranch feeds on sponges.

137. SAN DIEGO DORID

Diaulula sandiegensis

Identification: This doris usually has large brown to black ring-shaped markings on its back. These markings may also appear as solid blotches. The background color varies from white to pale yellow or pale brown. *Size*: Length to 3.25 inches (80 mm). *Range and Habitat*: Kachemak Bay, Alaska and Aluetian Islands to Cabo San Lucas, Baja California. *Diaulula* has also been recorded from Japan and Peter the Great Bay, Russia. Usually on hard bottoms to depths of less than 100 ft. (28 m). *Natural History*: This dorid feeds on sponges.

138. TOCHNI

Tochuina tetraquetra

Identification: This very large, orange nudibranch has white-tipped tubercles. The outer margin of the body has a white band. *Size*: Length to about 12 inches (30 cm). *Range and Habitat*: Kuril Islands, Russia to Alaska and Malibu, California. On hard bottoms to depths of at least 100 ft. (28 m). *Natural History*: Tochni feed on the soft coral *Gersemia* (# 81) and the gorgonian, *Lophogorgia* (# 84).

139. RAINBOW DENDRONOTUS

Dendronotus iris

Identification: This distinctive nudibranch varies highly in color, from red to orange, to grey and even white. It is distinguished by a white line along the edge of its foot. *Size*: Length to 8 inches (20 cm). *Range and Habitat*: Unalaska Island, Alaska to Cabo San Lucas, Baja California, Usually on soft bottoms to depths of about 75 ft. (20 m). *Natural History*: Rainbow *Dendronotus* feeds on tube-dwelling anemones (# 79). They also attach their egg masses to the parchment-like tubes. They can move around by swimming.

140. SPANISH SHAWL

Flabellina iodinea

Identification: This brightly colored animal has a purple body and orange cerata. The rhinophores are maroon. *Size*: Length to 1.75 inches (40 mm). *Range and Habitat*: Vancouver Island, British Columbia to San Benitos Islands, Baja California on Rocky areas to about 120 ft. (33 m). *Natural History*: Like the rainbow *Dendronotus*, this nudibranch can move about by swimming.

141. WHITE-LINED DIRONA

Dirona albolineata

Identification: The spear-shaped, large cerata have white borders, as does the tail and frontal veil. Body color is grey-white, salmon or purple. *Size*: Length to 7 inches (18 cm). *Range and Habitat*: Kachemak Bay, Alaska to San Diego, California. On rocky bottoms to depths of less than 100 ft. (28 m).

142. HERMISSENDA

Hermissenda crassicornis

Identification: The color patterns on the grey body; a gold line running down the center of the back, with blue bands on either side, and the white tipped brown cerata are distinctive. *Size*: Length to 2 inches (50 mm). *Range and Habitat*: Kodiak Island, Alaska to Punta Eugenia, Baja California and also in the northern part of the Sea of Cortez. Found on hard and soft bottoms to depths of less than 100 ft. (28 m).

143. HILTON'S AEOLID

Phidiana hiltoni

Identification: This common nudibranch has a white body with white-tipped, brownish cerata with a red line across the head extending onto the two oral tentacles. *Size*: Length to 2 inches (50 mm). *Range and Habitat*: Duxbury Reef, Marin County, California to Cedros Island, Baja California. On hard bottoms in depths less than 100 ft. (28 m). *Natural History*: *Phidiana* feeds on hydroids and other aeolid nudibranchs.

CLASS BIVALVIA
Scallops and Clams

144. ROCK SCALLOP
Crassedoma giganteum
(Formerly ***Hinnites giganteus***)
Identification: The juveniles, up to about 2 inches (50 mm), are not attached and have ribbed upper shells. Adults attach to substrate and have very thick valves with prominent ribs that have short spines. *Size:* Shell diameter to about 8 inches (200 mm). *Range and Habitat:* Queen Charlotte Islands, British Columbia to Punta Abreojos, Baja California. On rocks; low intertidal to about 150 ft (45 m).

145. ABALONE JINGLE
Pododesmus cepio
Identification: The rounded, very flat shell is attached on one side to rocks and other hard substrates. The valves are unequal in size, the lower, attached valve is the smaller of the two. *Size:* Diameter of shell about 3.5 inches (90 mm). *Range and Habitat:* Southern Alaska to Cabo San Lucas, Baja California. Attached to rocks, pier pilings, and docks, and even red abalone (# 112). Low intertidal to about 100 ft (30 m).

146. PACIFIC PINK SCALLOP
Chlamys rubida
Identification: This swimming scallop is almost round; the outer shell is ribbed and usually covered with a sponge. The sides of the hinge ("ears") are unequal in size. *Chlamys rubida* can be distinguished from *C. hastata* (# 147) by the lack of large spines on the ribs. *Size:* Diameter of shell to about 2.75 inches (70 mm). *Range and Habitat:* Southern Alaska to southern California. On rocky as well as soft bottoms; in subtidal depths to about 5244 ft (1600 m).

Note: The two species of *Chlamys* are indistinguishable underwater. The photos could be either species.

147. SPINY SCALLOP

Chlamys hastata

Identification: Differs from *Chalmys rubida* (# 146) in having large spines on the ribs. *Size:* Shell diameter to about 3.25 inches (82 mm). *Range and Habitat:* Aleutian Islands to southern California. On rocky as well as soft bottoms; in subtidal depths to about 120 ft (37 m).

148. GEODUCK CLAM

Panope generosa

Identification: The siphon of these large, edible clams is all that divers usually see. It is larger than the shell and appears as a smooth tube that protrudes above the sand or mud. *Size*: Length of shells to about 7 inches (175 mm). Length of siphon to about 40 inches (1m). *Range and Habitat:* Forrester Island, Alaska, to Scammon's Lagoon, Baja California. In sandy and mud bottoms, from low intertidal zone to 50 ft (15 m).

149. HORSE NECK CLAM

Tresus nuttalli

Identification: The tip of siphon is covered by two leather-like plates. When pulled together, they seal the opening. This characteristic separates this gaper clam from its look-alike *Tresus capa* (not illustrated). The shell is oblong and is not large enough to contain the retracted siphon. *Size:* Length of shell to about 8 inches (200 mm). *Range and Habitat:* Southern Alaska to Scammon's Lagoon, Baja California. Burrows in sand and mud bottoms; low intertidal to about 100 ft (30 m).

150. WART-NECK PIDDOCK

Chaceia ovoidea

Identification: Piddocks are boring clams that burrow into siltstone or sandstone reefs. The siphons are the only part of a living piddock you will ever see. The wart-necked piddock has a very distinctive dark brown siphon that extends above the rocks as much as 3 inches (75 mm). *Size*: Length of shell to about 4.5 inches (115 mm). *Range and Habitat:* Santa Cruz, California, to Bahia San Bartholome, Baja California, in bays and outer coastal waters. Very low intertidal to about 60 feet (18 m).

151. SCALESIDE PIDDOCK

Parapholas californica

Identification: The united siphons of this piddock usually do not protrude above the surface as in the wart-neck piddock (# 150). The color of the siphons is white with reddish-brown spots and blotches. *Size:* Length of shell to about 6 inches (150 mm). *Range and Habitat:* Bodega Bay, California, to Bahia San Bartolome, Baja California. In shale and clay reefs in exposed bays and outer coast. Low intertidal to about 60 ft (18 m).

CLASS CEPHALOPODA

Squid and Octopus

152. STUBBY SQUID

Rossia pacifica

Identification: The short body of this squid easily separates it from the market squid (# 153). *Size*: Length to about 5 inches (125 mm). *Range and Habitat:* Japan to southern California. On soft bottoms, occasionally observed in tidepools at night; subtidal depths 50 to 1200 ft (15 to 360 m).

Bernie Hanby

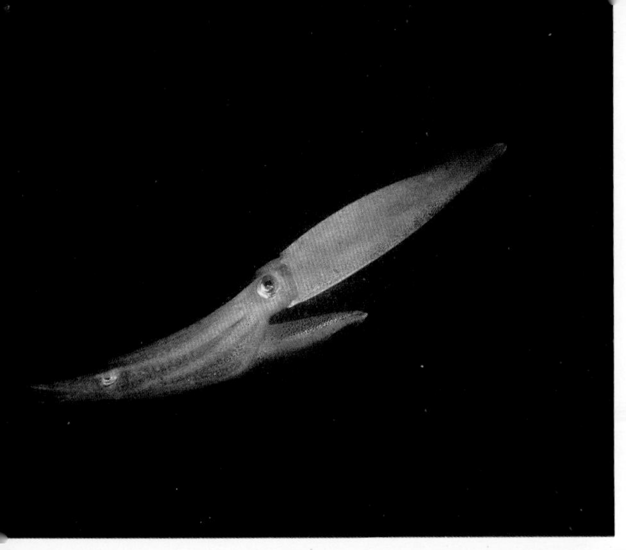

153. MARKET SQUID

Loligo opalescens

Identification: The mantle of this squid is not fused to the head and it is 4 to 5 times as long as wide. Color is variable; turns to brown or dark brown when animal is excited or frightened. *Size:* Length to about 11 inches (275 mm). *Range and Habitat:* Southern Alaska to Isla Guadalupe and Bahia Asuncion, Baja California. Pelagic; spawns over sand or mud bottoms. *Natural History:* The most common and abundant nearshore squid found off California. Adults feed on euphausiids, mysids, fishes, and even their own young.

154. GIANT PACIFIC OCTOPUS

Enteroctopus dofleini

Identification: This very large octopus has arms that are three to five times the length of the body. Color varies with background and mood, but is usually a reddish-brown. *Size:* Arm spread to about 10 ft (3 m); weight to about 100 Ib (45 kg). *Range and Habitat:* Northern Asia to California, also in southern hemisphere. On rocky as well as soft bottoms near a home crevice or cave; low intertidal to about 588 ft (180m).

155. RED OCTOPUS

Octopus rubescens

Identification: This common, small octopus has an ovoid body. The color varies from dull red to mottled white. There are often cirri on the skin. The arms are about four times the length of the body. *Size:* Length of body and arms to about 16 inches (45 cm). *Range and Habitat:* Alaska to Scammon's Lagoon, Baja California. Low intertidal to about 700 ft (210 m) on rocky and soft bottoms.

156. TWO-SPOT OCTOPUS

Octopus bimaculatus

Identification: This octopus has two eye spots (ocelli), one below each eye. Cirri are abundant on the skin. The arms are four to five times as large as the body. Color variable. *Size:* Length of body and arms to about 24 inches (61 cm). *Range and Habitat:* Santa Barbara, California, to Cabo San Lucas, Baja California and Sea of Cortez. In crevices and caves; low intertidal to about 165 ft (50 m). *Natural History:* Females brood the eggs for the 2 to 4 months required for hatching.

PHYLUM ARTHROPODA

Barnacles, Mantis Shrimp, Shrimps, Crabs, Hermit Crabs and Lobsters

Arthropods, which includes the insects, have evolved an external, segmented skeleton (exoskeleton). They can be found in all of the world's marine, freshwater, and terrestrial habitats. Most arthropods are mobile, but one group, the barnacles, after passing through a mobile larval stage, settle to the bottom and remain attached.

The arthropod body is typically divided into a head, a thorax, and an abdomen. The entire body, the antennae, and the legs are all segmented. All of the organ systems of higher animals are well developed in the arthropods, including eyes, nervous system, circulatory system, and excretory system. Their food ranges from tiny plankton to other small and large invertebrates or fishes. Many obtain nourishment by scavenging, but some arthropods are parasites. All members of this phylum reproduce sexually. A number of species of shrimp are protandric hermaphrodites, i.e., they first function as males, then after a year or two, transform into females.

Marine arthropods provide food for a host of predators including other arthropods and invertebrates, fishes, birds, reptiles, and mammals. Many species of crabs and shrimps support large commercial fisheries throughout the world.

At least 50 species of barnacles, and 300 species of shrimps, crabs, and lobsters have been recorded from this area.

CLASS CRUSTACEA
Subclass Cirripedia (Barnacles)

157. LEAF BARNACLE
Pollicipes polymerus

Identification: The base of the stalk consists of whorls of imbricate scales. Body enclosed by more than five plates. *Size:* Stalk length to about 3.5 inches (80 mm). *Range and Habitat:* Southern Alaska to at least Punta Abreojos, Baja California. Attached to rocks; intertidal to shallow subtidal.

158. GIANT ACORN BARNACLE
Balanus nubilus

Identification: This very large, unstalked, sessile barnacle is not easily confused with any other species; it has a large aperture and lacks longitudinal striations on any of the large body plates. *Size:* Diameter to 4.25 inches (110 mm). Height to about 3 inches (75 mm). *Range and Habitat:* Southern Alaska to La Jolla, California. Attached to rocks and pier pilings; low intertidal zone to depths of at least 300 ft (90m).

159. CALIFORNIA BARNACLE
Megabalanus californicus

Identification: The longitudinal red and white stripes on each plate are distinctive. *Size:* Diameter to about 2.5 inches (60 mm). Height to about 2 inches (50 mm). *Range and Habitat:* Humboldt Bay, California to Guaymas, Mexico. Attached to rocks, pilings, kelp, mussels, crabs and other hard shelled animals. Low intertidal to about 40 ft. (12 m).

Subclass Malacostraca
Order Stomatopoda
Mantis Shrimp

Bruce Bell

160. MANTIS SHRIMP

Hemisquilla ensigera californiensis

Identification: This elongate, shrimp-like crustacean is distinguished from the other species in this area by the bright blue walking legs and pliopods. *Size*: Length to about 12 inches (300 mm). *Range and Habitat:* Point Conception, California, to the Gulf of California and Panama. Burrows in soft bottoms; shallow subtidal to about 300 ft (90 m). *Natural History:* These animals should be handled with care as they are capable of inflicting severe cuts with the modified, folded back maxilliped (modified mouth parts).

Order Decapoda
Shrimp and Crabs

161. COONSTRIPE SHRIMP

Pandalus gurneyi

Identification: The body is a translucent brown to green with dark brown to black spots. On the abdomen and thin white lines on the anterior of the carapace. The rostrum is slightly longer than the carapace. *Size:* Length to about 6 inches (150 mm). *Range and Habitat:* Central California to southern California. On soft bottoms and rocky bottoms with crevices; intertidal depths to about 100 ft (30 m).

162. SPOT PRAWN

Pandalus platycerus

Identification: The color of this large shrimp is a light to dark translucent orange with a pair of distinctive white spots on the abdomen. *Size:* Length to about 10.5 inches (270 mm). *Range and Habitat:* Alaska to southern California. Usually on soft bottoms or rough rocky areas in steep canyons. Intertidal (in northern waters) to 1600 ft (500 m).

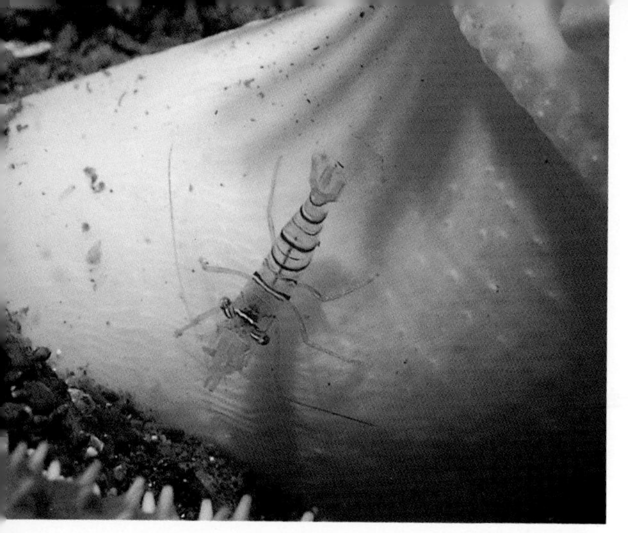

163. CLOWN SHRIMP

Lebbeus grandimanus

Identification: This distinctive, transparent shrimp has alternative bands of yellow and red on the body. *Size:* Length to about 2 inches (50 mm). *Range and Habitat:* Aleutian Islands to Puget Sound. Usually on rocks in association with the anemone *Cribrinopsis fernaldi* (# 58), in depths from 30 to about 100 ft (10 to 30 m).

Randy Morse

164. RED ROCK SHRIMP

Lysmata californica

Identification: The short rostrum and the reddish body, covered with darker bands and stripes, are distinctive. *Size:* Length to 2.75 inches (70 mm). *Range and Habitat:* Santa Barbara, California to Bahia Sebastian Vizcaino, Baja California. In rocky crevices during the day and on sand at night; shallow subtidal to depths of about 200 ft (60 m). *Natural History:* These shrimp are "cleaners." The California moray, garibaldi, and California spiny lobster regularly use their services.

165. CALIFORNIA SPINY LOBSTER

Panulirus interruptus

Identification: This distinctive, large, crustacean has very long antennae covered with small sharp spines. It lacks the large claws (chelae) of the Atlantic lobster. The front of the body, especially around the eyes, is covered with sharp spines. *Size:* Length to at least 2 ft. (60 cm). *Range and Habitat:* San Luis Obispo County,California to Rosalia Bay, Baja California. In crevices during daytime, forages on reef and sand bottom at night. Intertidal to about 200 ft (60 m).

166. HERMIT CRAB

Pagurus hemphilli

Identification: The dark red color of the legs and claws is distinctive. The legs and claws lack hairs. *Size:* Carapace length to about 0.5 inches (14 mm). *Range and Habitat:* Southern Alaska to southern California. On rocks in shallow waters. *Natural History:* These hermit crabs usually occupy the shells of the snails *Lithopoma* and *Tegula.*

167. HERMIT CRAB

Pagurus armatus

Identification: The size of the chelae is unequal; the right claw is larger on this large hermit crab. Color is usually orange with white bands. Legs and claws have spines and short claws. *Size:* Carapace length to about 0.75 inches (19 mm). *Range and Habitat:* Southern Alaska to southern California. Usually on soft bottoms in shallow water. *Natural History:* Commonly found in the shells of moon snails (# 124).

168. HERMIT CRAB

Elassochirus tenuimanus

Identification: The right claw (chela) is larger than the left claw and there are spines on the upper surface of the claws. Color is orange-brown. *Size:* Carapace length to about 1.25 inches (30 mm). *Range and Habitat:* Alaska to Washington. On rocks; intertidal to about 700 ft (210 m).

169. HERMIT CRAB

Paguristes ulreyi

Identification: The very hairy legs and claws (chelae) are distinctive. *Size*: Carapace length to about 1 inch (25 mm). *Range and Habitat:* British Columbia to Baja California. On rocks and soft bottoms to depths of about 50 ft (15 m). *Natural History:* Commonly found in the shells of snails *Lithopoma* (#120 and 121), *Polinices* (#124), and *Kelletia* (# 128).

Bernie Hanby

170. LITHOID CRAB

Acantholithodes hipsidus

Identification: This lithoid crab has short spines on the triangular-shaped carapace, as well as on the legs. *Size:* Carapace width to about 2 inches (50 mm). *Range and Habitat:* Bering Sea and Aleutian Islands to central California. On rocky bottoms; in shallow subtidal to about 500 ft (150 m).

171. BUTTERFLY CRAB

Cryptolithodes typicus

Identification: The rostrum narrows at the distal end and there are tubercles on the chelae. Color varies widely from white to red. *Size:* Width of carapace to about 3 inches (75 mm). *Range and Habitat:* Alaska to southern California. Uncommon in California. On rocks; from low intertidal to depths of about 240 ft (73 m).

172. UMBRELLA CRAB

Cryptolithodes sitchensis

Identification: The rostrum widens at the distal end and the smooth chelae distinguishes this umbrella crab from its look alike *C. typicus* (# 171). Color extremely variable. *Size:* Width of carapace to about 3 inches (75 mm). *Range and Habitat:* Alaska to San Diego, California. On rocks among attached coralline algae. Low intertidal to depths of at least 50 ft (15 m).

173. HAIRY CRAB

Hapalogaster cavicauda

Identification: The smooth carapace and legs of this crab are covered with short, brownish hair. *Size:* Carapace width to about 0.75 inch (19 mm). *Range and Habitat:* Cape Mendocino, California to Isla San Geronimo, Baja California. On rocky substrate; low intertidal to depths of about 50 ft (15 m). *Natural History:* This fuzzy crab obtains food by straining food particles from the water using long hairs on the outer mouth parts.

174. PUGET SOUND KING CRAB

Lopholithodes mandtii

Identification: These distinctive crabs have only three pairs of walking legs. The adult carapace is heavy, with four humps, and many blunt spines on the anterior edge. The claws are massive, while the legs are short and can be folded under the carapace. Juveniles are brilliantly colored and have an anterior bump that is larger than the other three bumps. This crab differs from its look alike *L. foraminatus* (not illustrated) by the lack of the semicircular indentation on the outer edge of each claw and on the forward edge of the front part of

juvenile

adult

legs. These indentations form a halo (foramen) when the claws and legs are tucked in. *Size:* Carapace width to about 9 inches (225 mm). Carapace width of *L. foraminus* to about 6 inches (150 mm). *Range and Habitat:* Kodiak Island, Alaska, to central California. On rocks as well as soft bottoms; shallow water to depths of 330 ft (100 m). *Lopholithodes foraminatus* ranges from Kodiak Island to southern California, in depths of 60 to about 2000 ft (20 to 600 m).

175. KING CRAB

Paralithodes camtschaticus

Identification: As with other lithoid crabs, the king crab has only three pairs of walking legs. The legs and carapace are covered with short, sharp spines. *Size:* Carapace width to about 12 inches (300 mm). *Range and Habitat:* Northwestern Pacific to Alaska and British Columbia. On soft bottoms, in depths of 10 to 600 ft (3 to 180 m). *Natural History:* The most valuable commercial crabs on the Pacific Coast of North America. These very large crabs live to at least 30 years. They feed on clams, sea urchins, and fish.

176. LITHOID CRAB

Phyllolithodes papillosus

Identification: This lithoid crab has a triangular-shaped carapace. The center of the carapace is depressed and surrounded by rounded tubercles. *Size:* Width of carapace to about 3 inches (75 mm). *Range and Habitat:* Alaska to southern California. Usually on rocky substrate; in shallow waters to about 100 ft (30m).

177. PELAGIC RED CRAB

Pleuroncodes planipes

Identification: This lobster-like pelagic crab is distinguished by its very long chelae and its short abdomen and tail. *Size:* Body length to about 3 inches (75 mm). *Range and Habitat:* Eureka, California (during an El Nino) to lower Baja California. Usually near the surface, but occasionally on or near bottom. *Natural History:* This very abun-dant crab is one of the most important food sources for midwater and surface fishes off Baja California.

Bernie Hanby

178. SQUAT LOBSTER

Munida quadrispina

Identification: This crustacean is very similar in appearance to the pelagic red crab (# 177). This crab is orange, rather than red. *Size:* Total length to about 5 inches (125 mm). *Range and Habitat:* Yakutat, Alaska to southern California. On soft bottoms in subtidal depths.

Marc Chamberlain

179. PURPLE GLOBE CRAB

Randallia ornata

Identification: The globular shaped, smooth, white carapace, mottled with purple, is very distinctive. *Size:* Width of carapace to about 2 inches (50 mm). *Range and Habitat:* Monterey Bay, California to Baja California. On soft bottoms; intertidal to about 70 ft (21 m). *Natural History:* Purple globe crabs bury themselves in the sand during the day and come out to feed at night.

Greg Jensen

180. SNOW CRAB

Chionoecetes bairdi

Identification: This large crab has chelae (claws) that are shorter than the four pairs of walking legs. The oval-shaped carapace has scattered spines and tubercles. *Size:* Carapace width to about 6 inches (150 mm). *Range and Habitat:* Bering Sea to Washington. On soft bottoms in waters as shallow as 30 ft (9 m) to depths of about 1500 ft (450 m). *Natural History:* These crabs are a very important part of the Pacific coast's commercial crab catch. They may live as long as 14 years.

181. MASKING CRAB

Loxorhynchus crispatus

Identification: The masking crab can be separated from the larger sheep crab by the coat of short hairs on the angular-shaped carapace. The carapace also has fewer tubercles. Chelipeds of males are longer than those of females. *Size:* Carapace width to about 3.5 inches (90 mm). *Range and Habitat:* Vancouver Island, British Columbia to Baja California. On rocky as well as soft bottoms; low intertidal to depths of about 600 ft (180 m).

182. SHEEP CRAB

Loxorhynchus grandis

Identification: This very large spider crab has a distinctive, robust, oval carapace covered with spines and tubercles. The males have much longer chelipeds (claws) than the females. *Size:* Carapace width to about 6.25 inches (159 mm). *Range and Habitat:* Point Reyes, California to Punta San Bartholeme, Baja California. On rocky as well as soft bottoms; depth range from 20 to about 500 ft (150 m). *Natural History:* The sheep crab does not use other organisms to mask its appearance.

183. SHARP-NOSED CRAB

Scyra acutifrons

Identification: The prominent two flat horns of the rostrum are broad and rounded on the sides. This crab is easily confused with the young of *Loxorhynchus crispatus* (# 181). *Size:* Carapace width to about 1.5 inches (38 mm). *Range and Habitat:* Alaska to Punta San Carlos, Baja California. On rocks, among algae and sessile animals; intertidal to about 300 ft (90 m).

184. DECORATOR CRAB

Oregonia gracilis

Identification: This spider crab is distinguished by its very long, smooth, and slender legs. The carapace is almost triangular. *Size:* Carapace width to about 3 inches (75 mm). *Range and Habitat:* Bering Strait to central California. On rocky bottoms, in depths from low intertidal to about 1300 ft (400 m). *Natural History:* This masking crab also uses sponges and hydroids as well as algae to camouflage itself.

185. MIMICKING CRAB

Miniulus foliatus

Identification: The lateral expansions of the smooth carapace suggest a leaf-like appearance. The smooth carapace often has attached sponges and bryozoans. *Size*: Carapace width to about 3 inches (75 mm). *Range and Habitat:* Alaska to San Diego, California. Usually on rocks in kelp areas; low intertidal to depths of about 400 ft (120 m). *Natural History:* These crabs feed on algae, primarily giant kelp, *Macrocystis.* They are part of the diet of several species of fishes.

Dale Glantz

186. SOUTHERN KELP CRAB

Taliepus nuttallii

Identification: The more convex carapace and prominent rostrum with a triangular notch separates this crab from *Pugettia producta* (# 188). Adults are usually bright red. *Size:* Carapace width to about 4 inches (100 mm). *Range and Habitat:* Santa Barbara, California, to Bahia Magdalena, Baja California. On rocky shores and bottoms in kelp forests; low intertidal to about 300 ft (90 m).

187. KELP CRAB

Pugettia richii

Identification: The carapace has four lateral, spiny projections. The eye stalk is located close to the second lateral projection from the anterior. The eye stalks of the similar species, *Pugettia gracilis* (not illustrated), are located close to the third lateral projection. The legs are relatively large and slender. *Size:* Carapace width to about 1.5 inches (30 mm). *Range and Habitat:* Prince of Wales Island, Alaska, to Isla San Geronimo, Baja California. *Pugettia gracilis,* Aleutian Islands to Monterey Bay, California. On rocky substrate, low intertidal to about 325 ft (98 m).

188. KELP CRAB

Pugettia producta

Identification: The relatively smooth carapace is usually brown to reddish in living animals. *Size:* Carapace widths 4.75 inches (93 mm). *Range and Habitat:* Yakutat, Alaska, to Punta Asuncion, Baja California. On rocky substrate and on giant kelp; low intertidal to about 250 ft (75 m). *Natural History:* This herbivore feeds on brown algae. Females are capable of producing offspring as often as every 30 days and may carry as many as 61,000 developing eggs.

189. ELBOW CRAB

Heterocrypta occidentalis

Identification: This distinctive crab has a much wider than long, wing-shaped carapace and very long, elbow-bearing arms with small chelae. *Size:* Carapace width to about 2 inches (50 mm). *Range and Habitat:* Central California to Baja California. On sand bottoms; intertidal and shallow subtidal to about 70 ft (21 m).

190. HELMET CRAB

Telmessus cheiragonus

Identification: The almost pentagonal shaped carapace is about 1/4 greater in width than length. The carapace, as well as chelae and legs are covered with bristles. *Size*: Carapace width to about 4 inches (100 mm). *Range and Habitat:* Pacific northwest to northern California. On rocky as well as sandy bottoms; low intertidal to about 130 ft (39m).

191. BROWN ROCK CRAB

Cancer antennarius

Identification: This cancer crab has a smooth light to dark-reddish carapace and dark spots on the underside of body, legs, and chelipeds. The antennae are more than twice the length of the eye stalks. *Size:* Carapace width to about 6 inches (150mm). *Range and Habitat:* Coos Bay, Oregon, to Bahia Magdelena, Baja California. On rocky as well as soft bottoms; very low intertidal to depths of about 100 ft (30 m).

192. YELLOW CRAB

Cancer anthonyi

Identification: This crab is more yellow than the brown rock crab (# 191) and lacks the spotting on the underside of the legs. The antennae are short. *Size:* Carapace width to about 7 inches (175 mm). *Range and Habitat:* Humboldt Bay, California, to Bahia Magdalena, Baja California. Usually on soft bottoms; subtidal depths to about 140 ft (48 m). *Natural History:* The yellow crab is one of the most important crabs caught by commercial trappers in southern California. This crab has been observed cleaning parasites from sand bass, *Paralabrax nebulifer,* in southern California.

193. CANCER CRAB

Cancer branneri

Identification: Similar to *Cancer jordani* (# 195) in having a hairy carapace, legs, and chelipeds. The shell is bright orange-red. The most posterior tooth on the carapace is prominant and sharp. *Size:* Carapace width to about 3 inches (75 mm). *Range and Habitat:* Alaska to southern California. On rocky and soft bottoms; shallow subtidal depths to about 100 ft (30m).

194. SLENDER CRAB

Cancer gracilis

Identification: The slender crab can be distinguished from juvenile Dungeness crabs (# 196) by the purple coloration on the upper surface of the legs. *Size:* Carapace width to about 4 inches (100 mm). *Range and Habitat:* Alaska to Bahia Playa Maria, Baja California. On soft bottoms; subtidal depths to about 350 ft (105 m). *Natural History:* Slender crabs are fed on by starry flounders, *Platichthys stellatus.* The megalop larvae are often found on the purple jellyfish, *Chrysaora colorata* (# 47).

195. HAIRY CANCER CRAB

Cancer jordani

Identification: This hairy crab can be separated from the look alike *Cancer branneri* (# 193) by its most posterior tooth on the carapace, which is inconspicuous. *Size:* Carapace width to about 1.25 inches (33 mm). *Range and Habitat:* Coos Bay, Oregon, to Cabo Thurloe, Baja California. On rocky and soft bottoms; intertidal to about 350 ft (105 m).

196. DUNGENESS CRAB

Cancer magister

Identification: The relatively smooth carapace has 10 small teeth on each side. The posterior tooth is the largest and distinguishes this crab from the other *Cancer* crabs. *Size:* Carapace width to about 9 inches (225 mm). *Range and Habitat:* Aleutian Islands, Alaska, to Santa Barbara, California. On soft bottoms; from very low intertidal to about 1200 ft (360 m). *Natural History:* This is the most important commercially trapped crab on the Pacific Coast south of Alaska. Dungeness crabs are carnivorous feeding on a variety of small invertebrates and fishes. Large adults prey on the juvenile crabs during their first year of life.

197. OREGON CANCER CRAB

Cancer oregonensis

Identification: The carapace of this *Cancer* crab differs from all others of this genus by its more rounded shape; it lacks the angle where the anterior and posterior margins meet. The legs are hairy. *Size:* Carapace width to about 1.5 inches (40 mm). *Range and Habitat:* Pribilof Islands, Alaska to southern California. On holes and crevices on rocky substrate; from intertidal to about 1450 ft (435 m).

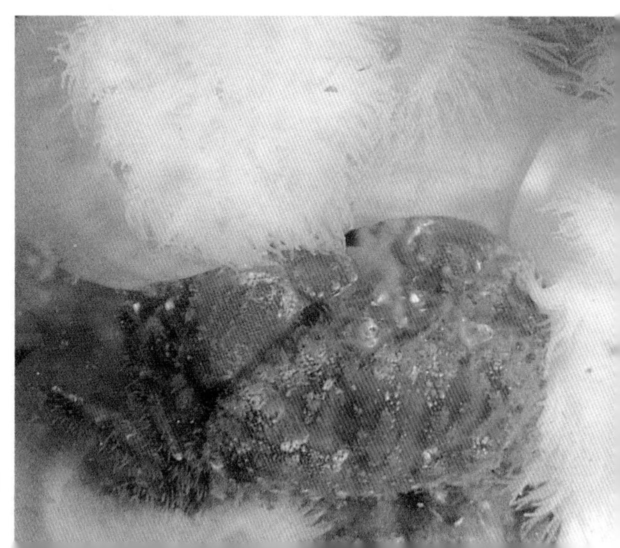

198. RED ROCK CRAB

Cancer productus

Identification: The bright red carapace, legs, and claws of the adult are distinctive. The carapace of juveniles has a wide variety of color patterns. *Size:* Carapace width to about 7 inches (175 mm). *Range and Habitat:* Alaska to San Diego, California. On rocky as well as soft bottoms; from the low intertidal (juveniles) to about 300 ft (90 m).

Marc Chamberlain

199. SWIMMING CRAB

Portunus xantusii

Identification: This relative of the Atlantic Coast soft shell crab can be distinguished from the other swimming crab that occurs on this coast, *Callinectes arcuatus* (not illustrated), in having a spine on the inner side of the chelae next to the last joint. *Size:* Carapace width to about 3 inches (75 mm). *Range and Habitat:* Morro Bay, California to Topolobampo, Mexico. Usually partially or fully buried in sand. Has been observed swimming on surface. Low intertidal to about 600 ft (180 m). *Callinectes arcuatus:* Anaheim Slough, California, to Peru. From low intertidal to about 90 ft (27 m).

Marc Chamberlain

Class Pycnogonida

200. SEA SPIDER

Rhynchothorax philopsammum

Identification: This sea spider is distinguished by the long legs and the lack of claws or pinchers (chelifores). *Size:* Length of body to about 0.5 inches (12 mm). *Range and Habitat:* Northern and southern California. On rocky bottoms, in shallow

PHYLUM BRYOZOA

Moss Animals

Moss animals are colonial invertebrates that occur in both marine and freshwater environments. These sessile animals are often confused with hydroids, corals, or, sometimes, marine algae. Each individual animal in a colony is called a zooid. A typical zooid consists of a body within a calcareous chamber (zooecium) with an exposed tentacle crown. Each zooid possesses digestive and reproductive systems and a simple nervous system. There are no advanced sensory organs. Bryozoans reproduce sexually, but growth of the colony is by asexual budding. Some species are hermophroditic. The tentacles are used to capture bacteria, phytoplankton and other small organisms and detritus suspended in the water. There are specialized zooids whose sole function is reproduction, protection of the colony, brooding of embryos, or to attach the colony to the substrate. Many other species of invertebrates prey on bryozoans. Approximately 300 species of moss animals have been described from the area covered by this field guide.

Class Gymnolaemata

201. FLUTED BRYOZOAN

Hippodiplosia insculpta

Identification: The colonies consist of fan-like folds that are curled. Color varies from light yellow to orange and tan. *Size:* Colonies to 4 inches (100 mm) in height and about 5 inches (125 mm) in width. *Range and Habitat:* Gulf of Alaska to Gulf of California and Isla de Coco, Costa Rica. On hard substrate; from the low intertidal to about 780 ft (234 m).

202. LACY BRYOZOAN

Phidolopora labiata

(Formerly ***Phidolopora pacifica)***

Identification: The upright colonies form a convoluted mass of lace-like structures. Color is usually orange. *Size:* Colonies about 8 inches (200 mm) in diameter and 4 inches (100 mm) in height. *Range and Habitat:* British Columbia to South America. On rocks, from the very low intertidal (shaded pools) to about 665 ft (200 m).

203. BRYOZOAN

Costazia costazi

Identification: The colony has a fuzzy appearance and forms upright, single or forked nodules. Color is usually red-orange. *Size:* Colony height to about 2 inches (50 mm). Width of colony to about 3 inches (75 mm). *Range and Habitat:* British Columbia to Baja California. On rocks and kelp fronds; from low intertidal to about 330 ft (100 m).

204. SOUTHERN STAGHORN BRYOZOAN

Diaperoecia californica

Identification: Colonies in coral-like masses. The cross section of individual branches is flattened. Color light to dark yellow. *Size:* Colony height to about 4 inches (100 mm), width to about 6 inches (150 mm). *Range and Habitat:* British Columbia to Costa Rica. On rocks and other hard substrates. Very low intertidal to about 615 ft (185 m).

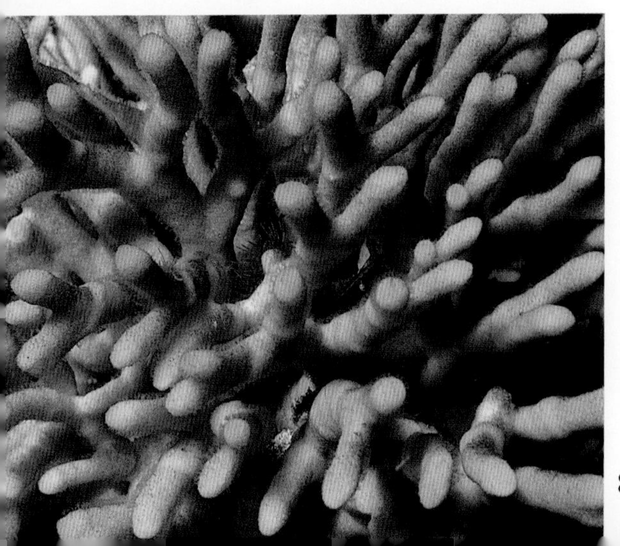

205. NORTHERN STAGHORN BRYOZOAN

Heteropora pacifica

Identification: The upright coral-like branches of the colony are round in cross section; color is light yellow. *Size:* Colony height to about 4 inches (100 mm); width to about 4 inches (150 mm). *Range and Habitat:* Alaska to Central California. On rocks; in subtidal depths to about 90 ft (27 m).

206. MOSS ANIMAL

Bugula californica

Identification: The colony is composed of fronds that consist of spiral whorls of branches. Color white to orange. *Size:* Colony height to about 2 inches (50 mm) width to about 4 inches (100 mm). *Range and Habitat:* British Columbia to Galapagos Islands. On rocks; from shallow subtidal to about 200 ft (60m).

207. BROWN BRYOZOAN

Bugula neritina

Identification: The reddish-brown or purplish brown bushy colonies are distinctive. *Size*: Height to about 4 inches (10 cm). *Range and Habitat*: Monterey Bay, California to Panama and Galapagos Islands. On giant kelp fronds, Pilings, boat hulls to depths of about 28 ft. (80 m). *Natural History*: Individuals (zooids) that are members of the colony are hermaphoditic. Colonies may only live for 1 or 2 years.

208. KELP LACE BRYOZOAN

Membranipora membranacea

Identification: Circular colonies in the form of flat encrustations of a single layer of zooids. The colonies have a reticulate honeycomb appearance. *Size*: Diameter to about 2 inches (5 cm). *Range and Habitat*: Alaska to Baja California. On giant kelp and other algae, low intertidal to depths of less than 100 ft. (28 m). *Natural History*: Because the kelp dies seasonally, colonies rarely survive for more than a year. This bryozoan provides food for three species of the nudibranch genus *Corambe*.

209. STICK BRYOZOAN

Microporina borealis

Identification: This branching colony is in the form of short, branching segments, elliptical in cross-section. *Size*: Height to about 3 inches (7. cm). *Range and Habitat*: Arctic region to Washington. On hard bottoms, subtidal depths to 132 ft. (400 m).

210. PURPLE BRYOZOAN

***Disporella* sp.**

Identification: This purple, encrusting bryozoan is often confused with the encrusting hydrocoral (#45), but lacks large pores containing polyps. *Size*: Width of colony to about 3 inches (78 mm). *Range and Habitat*: British Columbia to Baja California. On rocks and shells in depths of less than 100 ft. (28 m).

PHYLUM BRACHIOPODA

Lampshells

This small group of exclusively marine animals is often confused with scallops, clams, or mussels. All lampshells are sessile; the two shells are usually attached to the substrate by a stalk (pedicel). Brachiopods possess all of the organ systems of higher animals; however, the nervous system is not as complex and they lack complex sensory organs, such as eyes. All are filter feeders; they use appendages called lophophores to collect food as it passes by. They all reproduce sexually. Brachiopods are important to paleontologists, who use them to date sedimentary rocks. Sixteen species of lampshells have been recorded from southern Alaska to California.

211. LAMPSHELL

Terebratalia transversa

Identification: The lip of the shell is bent, not straight as in most bivalve molluscs. Also, the valves are dorsal and ventral rather than lateral. The valves can be smooth or have radial ribs. *Size:* Length to about 2 inches (50 mm). *Range and Habitat:* Alaska to Baja California. Attached to rocks; from low intertidal zone to about 6000 ft (1800 m). *Natural History:* This lampshell is known to have existed as a species for at least 20 million years. They are preyed upon by crabs.

PHYLUM PHORONIDA

Phoronids

All phoronids are marine and most species occur in temperate seas. Because of their general morphology, they are often confused with polychaete worms. The body of the animal is buried in sand or mud bottoms with the feeding tentacular crown exposed above the surface. A few species are found in hard substrate, such as shells, and a few are part of the encrusting community. While phoronids have well-developed circulatory, digestive, and reproductive systems, the nervous system is rather simple and they lack complex sensory organs.

Feeding is accomplished by the tentacles that capture food items that drift by. Reproduction in phoronids is entirely sexual. Only six species of phoronids have been reported from the ocean waters between southern Alaska and California.

CLASS PHORONIDA

212. CALIFORNIA PHORONID

Phoronopsis californica

Identification: The tentacles of the exposed portion are arranged in four to nine whorls on each side. Color ranges from orange to light yellow. *Size:* Height above sand to about 1.5 inches (40 mm). *Range and Habitat:* Point Conception, California to Newport Bay, California. Low intertidal; soft bottoms to about 115 ft (35 m).

Marc Chamberlain

213. VANCOUVER PHORONID
Phoronis vancouverenisis
Identification: These animals form colonies in the form of clumps or sheets. The color is usually white, red blood may be visible through the body wall. *Size*: Individual lengths to about 1.5 inches (40 mm). *Range and Habitat*: British Columbia to southern California. On hard substrate covered with sand, silt or mud. In depths from low intertidal to less than 100 ft. (28 m). *Natural History*: This species is hermaphroditic. The embryos are brooded; individuals release sperm packets into the surrounding waters, where adjacent animals can be fertilized internally. The larvae are pelagic.

PHYLUM ECHINODERMATA
Feather Stars, Sea Stars, Sea Urchins, Sea Cucumbers, and Brittle Stars

The echinoderms occur only in the marine environment. All of the species are mobile. These “spiny skinned” (literal translation of echinoderm) animals are unique in the construction of their endoskeleton; which is composed of large numbers of calcium carbonate crystals imbedded in the tissues. They are also unique in their possession of a system of water-filled tubes, which is the basis of their tube feet (podia). The tube feet are used to move about, cling to the substrate, for respiration, and to
capture food. Echinoderms have a nervous system, but lack advanced sensory organs such as eyes. They also have a complex digestive system. Sexes are separate in most echinoderms and fertilization usually takes place in the water column after the release of sperm and eggs.

In the area from Alaska to California, there are approximately 10 species of feather stars, 105 species of sea stars, 25 species of sea urchins, 80 species of sea cucumbers, and 65 species of brittle stars.

CLASS CRINOIDEA
Feather Stars

214. FEATHER STAR
Florometra serratissima
Identification: The 10 long arms have slender branches. Color is tan to reddish tan. *Size:* Height to about 10 inches (250 mm). *Range and Habitat:* Alaska to Baja California. On soft and hard bottoms; shallow subtidal (in northern part of range) to depths of about 3300 ft. (990 m).

CLASS HOLOTHUROIDEA
Sea Cucumbers

215. SEA CUCUMBER
Cucumaria piperata

Identification: This small sea cucumber has 10 branched tentacles. The color is usually white, with brown or black speckles. *Size:* Length to about 2.5 inches (60 mm). *Range and Habitat:* British Columbia to Baja California. In rocky areas, in crevices, from intertidal to about 275 ft (82 m).

216. ORANGE SEA CUCUMBER
Cucumaria miniata

Identification: This bright orange sea cucumber has 10 branched tentacles. There are five rows of tube feet. *Size:* Length to about 10 inches (250 mm). *Range and Habitat:* Gulf of Alaska, to Avila Beach, California. In rocky areas with crevices; from low intertidal zone to about 80 ft. (24 m). *Natural History:* The branched tentacles are used to collect drifting food. This sea cucumber is preyed upon by the sea star *Solaster stimpsoni* (# 242).

217. WHITE SEA CUCUMBER
Eupentacta quinquesemita

Identification: This sea cucumber can not completely retract its long tube feet. There are 10 yellow or white branched tentacles, 8 long and 2 short. *Size:* Length to about 4 inches (100 mm). *Range and Habitat:* Japan to Sitka, Alaska to Sacramento Reef, Baja California. On rocky substrate from the low intertidal to about 50 ft (15 m). *Natural History:* The tentacles usually remain retracted during daylight hours.

218. RED SEA CUCUMBER

Pachythyone rubra

Identification: The tube feet are not arranged in any patterns but are scattered over entire body. Color, orange to orange-red above, white below. *Size:* Length to about 3 inches (75 mm). *Range and Habitat:* Monterey Bay to southern California. On rocky and sandy bottoms, from the low intertidal to about 60 ft (18 m). *Natural History:* The females brood their eggs.

219. CALIFORNIA SEA CUCUMBER

Parastichopus californicus

Identification: This large sea cucumber has prominent, stiff, conical papillae. The tube feet are on the ventral side. Color is dark red, brown, or yellow. *Size:* Length to about 16 inches (40 cm). *Range and Habitat:* Alaska to Isla Cedros, Baja California. On rocks and soft bottoms; from the low intertidal to about 300 ft (90 m). *Natural History:* This is the largest sea cucumber in our area. It feeds on detritus and small organisms. Unfortunately, sea cucumbers are now the subject of a growing commercial fishery. This fishery developed before a scientific management plan was in place.

220. WARTY SEA CUCUMBER

Parastichopus parvimensis

Identification: This brownish sea cucumber is covered with small, black-tipped papillae. *Size:* Length to about 10 inches (250 mm). *Range and Habitat:* Monterey Bay, California to Isla Cedros, Baja California. On rocky and soft bottoms; from the low intertidal to about 100 ft (30 m).

221. SWEET POTATO

Caudina arenicola

Identification: This smooth sea cucumber lacks tube feet. Color is yellow, mottled with brown. *Size:* Length to about 10 inches (250 mm). *Range and Habitat:* Southern California to Baja California. On soft bottoms, in shallow subtidal depths. Eats sand and digests off the detritus.

222. SLIPPER SEA CUCUMBER

Psolus chitonoides

Identification: The bright red, branched tentacles extend out from a flattened "foot" that has calcareous plates covering the thick skin. There are three double rows of tube feet on the underside of the "foot." *Size:* Length to about 5 inches (125 mm). *Range and Habitat:* Pribilof Islands to Baja California. On rocks, from the intertidal to about 330 ft (100 m).

CLASS ECHINOIDEA

Sea Urchins

223. SLATE PENCIL URCHIN

Eucidaris thouarsii

Identification: The large, thick, blunt spines, often covered with encrusting animals, are distinctive. Color is reddish brown to purple. *Size:* Diameter to about 6 inches (150 mm). *Range and Habitat:* Santa Catalina Island, California, to Ecuador and the Galapagos Islands. Between or under rocks; from the low intertidal to about 450 ft (135 m).

224. CROWNED SEA URCHIN

Centrostephanus coronatus

Identification: The long, dark, sharp spines are usually at least three times as long as the diameter of the test. Color of spines is dark purple, sometimes with white bands. *Size:* Diameter of test and spines to about 7 inches (175 mm). *Range and Habitat:* Channel Islands, California to Galapagos Islands. On rocks, foraging at night on soft bottoms; from the low intertidal to about 110 ft (125m).

225. WHITE SEA URCHIN

Lytechinus anamesus

Identification: The sharp white spines are usually shorter than the diameter of the test. Color of test is white with dark blotches. *Size:* Diameter to about 3 inches (75 mm). *Range and Habitat:* Channel Islands, California, to Gulf of California. On soft as well as rocky bottoms; from the shallow subtidal to about 1000 ft (300 m). *Natural History:* Large aggregations of these algae feeders often gather at food sources.

226. PURPLE SEA URCHIN

Strongylocentrotus purpuratus

Identification: This distinctive sea urchin has short purple spines. Juveniles less than 1-inch in diameter and have spines that are greenish. *Size:* Diameter to about 3.5 inches (85 mm). *Range and Habitat:* Cook's Inlet, Alaska, to Isla Cedros, Baja California. On rocks; from the low intertidal to about 30 ft (10 m). *Natural History:* These sea urchins can excavate circular holes in sandstone and shale rocks, using their sharp spines and teeth. They feed on algae.

227. RED SEA URCHIN

Strongylocentrotus franciscanus

Identification: This large sea urchin has sharp spines that are about 1/2 as long as the diameter of the test. Color ranges from red, red brown, to dark purple. *Size:* Diameter to about 10 inches (250 mm). *Range and Habitat:* Northern Japan and Alaska to Isla Cedros, Baja California. On rocks; from the very low intertidal to about 300 ft (90 m). *Natural History:* This sea urchin supports a very large commercial fishery on the Pacific coast of North America. Divers collect them for their roe. Another predator is the sea otter.

228. GREEN SEA URCHIN

Strongylocentrotus droebachiensis

Identification: This northern species has short, light green spines. *Size:* Diameter to about 4 inches (100 mm). *Range and Habitat:* Circumpolar, on this coast south to Washington. On rocks, as well as soft bottoms; from the intertidal to about 426 ft (130 m). *Natural History:* Green sea urchins feed on algae. They are sought after as food by the predatory sea star, *Pycnopodia helianthoides* (# 256).

229. HEART URCHIN

Lovenia cordiformis

Identification: The oval-shaped test has very long spines on the upper surface. Color of the test varies from white, gray-brown, yellow, red, to purple. *Size:* Length of test to about 3 inches (75 mm). *Range and Habitat:* San Pedro, California, to Panama and Galapagos Islands. On and buried in soft bottoms; from the very low intertidal to about 460 ft (140 m).

Marc Chamberlain

230. SAND DOLLAR

Dendraster excentricus

Identification: This flat, irregular urchin has evenly developed, short, hair-like spines. Color varies from gray- lavender, brown, red brown, to dark purple. *Size:* Diameter to about 4 inches (100 cm). *Range and Habitat:* Alaska to Baja California. Partially to completely buried in sand; from the low intertidal to about 130 ft (40 m).

CLASS ASTEROIDEA

Sea stars

231. SAND STAR

Luidia foliolata

Identification: This star has five long arms that are flattened and rectangular in cross section and relatively smooth. Color is gray or brown. *Size:* Diameter to about 16 inches (41 cm). *Range and Habitat:* Southeast Alaska to San Diego, California. On soft bottoms; in subtidal depths to about 2,040 ft. (612 m). *Natual History:* Food of this star consists of bivalve molluscs, sea urchins, sea cucumbers, brittle stars, worms, and small crustaceans.

232. SPINY SAND STAR

Astropecten armatus / Astropecten verrilli

(These two may be synonymous)

Identification: This gray to tan-colored star has plates and spines on the edges of each arm. There are no suckers on the tube feet. *Size:* Diameter about 6 inches (150 mm). *Range and Habitat:* San Pedro, California to Ecuador. On soft bottoms, usually sand; from the very low intertidal to about 180 ft (54 m). *Natural History:* These sea stars are capable of moving on or below the surface of the sand. They feed on snails, such as the olive snail, *Olivella biplicata.*

233. COOKIE STAR

Ceramaster patagonicus

Identification: The five triangular-shaped arms are very short and have no spines on the upper surface, but there are large plates on the outer edge of each arm. Color is light orange to red-orange. *Size:* Diameter to about 8 inches (200 mm). *Range and Habitat:* Bering Sea to Mexico and Straits of Magellan, South America. On rocks, as well as soft bottoms; subtidal depths to about 785 ft (235 m).

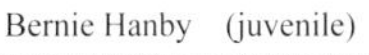

Bernie Hanby (juvenile)

234. SPINY RED STAR

Hippasteria spinosa

Identification: Similar to Red Sea Star (# 235) in shape but possesses tall, slender spines in upper surface and edges of arms. *Size:* Diameter to about 8 inches (200 mm). *Range and Habitat:* Kodiak Island, Alaska, to southern California. On soft bottoms; in subtidal depths to about 1700 ft (510 m). *Natural History:* These stars prey on sea pens.

(adult)

235. RED SEA STAR

Mediaster aequalis

Identification: This bright red star has five tapering arms and a broad central disc. The tube feet are red to flesh colored. *Size:* Diameter to about 8 inches (200 mm). *Range:* Alaskan Peninsula to Baja California. On rocks as well as soft bottoms; in subtidal depths to about 50 to 1650 ft (15 to 495 m). *Natural History:* These stars feed on sponges, bryozoans, sea pens, and also scavenge on dead animals.

236. BAT STAR

Asterina miniata (Formerly ***Patiria miniata***)

Identification: This very common sea star varies greatly in color. It lacks spines or pedicellariae. The number of arms is usually five, but occasionally four to nine. *Size:* Diameter to about 8 inches (200 mm). *Range and Habitat:* Sitka, Alaska to Isla Cedros, Baja California. On rocks as well as sandy bottoms; low intertidal to about 950 ft (285 m). *Natural History:* This distinctive star feeds on a variety of plant and animal matter, dead or alive. They feed by extending their stomachs over the food item.

237. SPINY SEA STAR

Poraniopsis inflata

Identification: This sea star has thick arms and prominent white spines on the upper surface and edges of the arms. Color is cream to orange. *Size:* Diameter to about 6 inches (150 mm). *Range and Habitat:* Japan and, on this coast, from Queen Charlotte Islands to San Diego, California. On rocks; in subtidal depths from 30 to 1220 ft (11 to 366 *m). Natural History:* Probably feeds on cup corals, such as *Paracyathus* (# 77).

238. LEATHER STAR

Dermasterias imbricata

Identification: The smooth, leathery-like skin of this sea star is distinctive. *Size:* Diameter to about 10 inches (250 mm). *Range and Habitat:* Prince William Sound, Alaska, to Sacramento Reef, Baja California. On rocks, occasionally on sand; from the very low intertidal to about 300 ft (90 m). *Natural History:* This star feeds on sea anemones, particularly *Epiactis prolifera* (# 57), as well as sea cucumbers and sea urchins.

239. ROSE STAR

Crossaster papposus

Identification: The disc diameter of this sea star is almost equal to the length of an individual arm. There are 8 to 14 arms. Clusters of spines cover the upper surface. *Size:* Diameter to about 10 inches (250 mm). *Range and Habitat:* Circumpolar. On this coast, south to Puget Sound, Washington. On soft as well as rocky bottoms. From the low intertidal to about 4000 ft (1200 m). *Natural History:* Rose stars feed on sea pens, brozoans, tunicates, and bivalve molluscs.

240. DAWSON'S SUN STAR

Solaster dawsoni

Identification: This 12- to 13-armed sea star varies in color from brown to gray-yellow, and occasionally red or orange. There are flat topped spinlets on the upper surface. *Size:* Diameter to about 20 inches (51 cm). *Range and Habitat:* Aleutian Islands to Point San Luis, California. On rocks; from the low intertidal to about 1400 ft (420 m). *Natural History:* This sun star preys on other sea stars particularly *Solaster stimpsoni* (# 242).

241. NORTHERN SUN STAR
Solaster endeca

Identification: The 7 to 13 short, tapering arms of this sea star are shorter than the other species of *Solaster.* The color is usually red or orange, occasionally with a purple stripe down each arm. *Size:* Diameter to about 16 inches (40 cm). *Range and Habitat:* Circumpolar. On this coast, to Puget Sound, Washington. On rocky as well as soft bottoms; from the intertidal to about 1580 ft (475 m). *Natural History:* This sun star is also a predator on other sea stars and sea cucumbers.

242. STIMPSON'S SUN STAR
Solaster stimpsoni

Identification: This sea star usually has 10 arms (range 9 to 11) that lack spines. The color varies from orange to red, with a blue-grey stripe along the upper surface of each arm. *Size:* Diameter to about 20 inches (51 cm). *Range and Habitat:* Bering Sea to Salt Point (Sonoma County) California. On rocky bottoms; from the very low intertidal to about 2000 ft (600 m). *Natural History:* This carnivorous star feeds on sea cucumbers, tunicates, lampshells, and sea pens.

Bernie Hanby

243. WRINKLED SEA STAR
Pteraster militaris

Identification: This star is similar to *P. tesselatus* (# 244), but it has longer, more slender arms. The upper surface (aboral) is wrinkled. Color is usually pale yellow. *Size:* Diameter to about 6 inches (150 mm). *Range and Habitat:* Circumpolar. In the Pacific, from Japan to Bering Sea and Oregon. On soft bottoms; in subtidal depths 30 to about 3600 ft (10 to 1100 m). *Natural History:* The females brood large, yolky eggs.

244. CUSHION STAR

Pteraster tesselatus

Identification: This star has stubby, broad, thick arms and lacks spines. The osculum is in the center. Color varies from tan to cream to yellow or dull gray. *Size:* Diameter to 6 inches (150 mm). *Range and Habitat:* Japan and Bering Sea to Carmel Bay, California. On rocks; in subtidal depths of 30 to 1450 ft (9 to 435 m). *Natural History:* Cushion stars feed primarily on sponges.

245. SEA STAR

Henricia aspersa

Identification: The color of this star is usually a brownish-red. It differs from the Blood Star (# 246) by having an open mesh-work of ridges on the uppper surface. Each ridge has groups of small spines. *Size:* Diameter to about 12 inches (300 mm). *Range and Habitat:* Sea of Japan and Bering Sea south to Santa Barbara, California. On rocks as well as soft bottoms; in subtidal depths to 3000 ft (900 m).

246. BLOOD STAR

Henricia leviuscula

Identification: The long, tapering arms are almost round in cross section. Usually with five arms, but occasionally four or six. They lack pedicellariae. Color varies from orange to red, often banded with darker shades. *Size:* Diameter to about 8 inches (200 mm). *Range and Habitat:* Aleutian Islands to Bahia Tortugas, Baja California. On rocks, particularly those encrusted with sponges and bryozoans; from the low intertidal to about 1330 ft (400 m).

247. SEA STAR

Henricia sanguinolenta

Identification: The long, slender arms of this star are thickened at the base and there is usually a crease between the arms extending almost to the center of the disc. Color ranges from nearly white to orange. *Size:* Diameter to about 6 inches (150 mm). *Range and Habitat:* Circumpolar. On this coast south to Washington. On rocks as well as soft bottoms; in subtidal depths to about 8000 ft (2400 m). *Natural History:* The females brood their eggs.

248. FRAGILE RAINBOW STAR

Astrometis sertulifera

Identification: The five long arms of this sea star have separated tapering spines on the upper surface. Each spine is encircled by pedicellariae. Color of the upper body is brown or green. The spines are orange. *Size:* Diameter to about 10 inches (250 mm). *Range and Habitat:* Santa Barbara, California to Gulf of California. On rocky bottoms; from the very low intertidal to about 130 ft (40 m). *Natural History:* This star's pedicellariae are capable of capturing mobile prey, even fish.

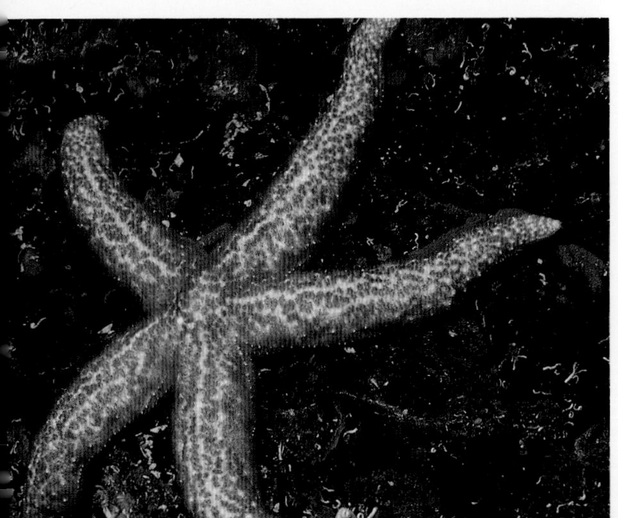

249. FALSE OCHRE STAR

Evasterias troschelii

Identification: The small disc and slim tapering arms distinguish this star from the ochre star (# 254); the spines on the aboral surface do not form a stellar pattern. Color varies from orange to brown and blue gray. *Size:* Diameter to about 12 inches (300 mm). *Range and Habitat:* Pribilof Islands to Carmel Bay, California. On rocks as well as soft bottoms; from the low intertidal to about 230 ft (69 m).

250. SIX-ARM SEA STAR

Leptasterias hexactis

Identification: This small, six armed sea star can be confused with small blood stars, (# 246), but *Leptasterias* arms are much shorter and more flattened in cross-section. The aboral surface is covered with small spines and pedicellariae. Color is black, brown, red, or greenish. *Size:* Diameter to about 3 inches (75 mm). *Range and Habitat:* San Juan Islands, Washington, to Santa Catalina Island, California. On rocks; from the intertidal to shallow subtidal.

251. RAINBOW STAR

Orthasterias koehleri

Identification: This colorful star has prominent, sharp spines on the arms. The color varies from rose-pink mottled with gray to red mottled with yellow. *Size*: Diameter to about 15 inches (38 cm). *Range and Habitat:* Yakutat Bay, Alaska, to southern California. On rocky as well as soft bottoms; from the very low intertidal to about 800 ft (240 m). *Natural History:* Molluscs form the major portion of the diet of this attractive sea star.

252. SHORT SPINED SEA STAR

Pisaster brevispinus

Identification: The aboral spines are very short, as compared with *Pisaster ochraceus* (# 254) and *P. giganteus* (# 253). The color is always pink. *Size:* Diameter to about 24 inches (60 cm). *Range and Habitat:* Sitka, Alaska to La Jolla, California. On rocky and soft bottoms; from the low intertidal to about 300 ft (90 m). More common in bays than on the open coast. *Natural History:* Clams, snails, and sand dollars are the prey of this very large sea star. They also scavenge on dead fish and squid.

253. GIANT SPINED STAR

Pisaster giganteus

Identification: The long spines are uniformly spaced and have swollen tips. Each spine is surrounded by a circle of blue. *Size:* Diameter to about 22 inches (56 cm). *Range and Habitat:* Vancouver Island, British Columbia to Isla Cedros, Baja California. On rocky as well as sand bottoms; from the very low intertidal to about 300 ft (90 m). *Natural History:* This sea star prefers mussels for food, as well as other molluscs and barnacles.

254. OCHRE STAR

Pisaster ochraceus

Identification: This very common, intertidal sea star usually has five thick arms, but the number varies from four to seven. There are numerous small white spines on the aboral surface arranged in a reticular pattern. Color varies from gray to orange. *Size:* Diameter to about 14 inches (35 cm). *Range and Habitat:* Prince William Sound, Alaska to Isla Cedros, Baja California. On rocky bottoms; from the intertidal to about 300 ft (90 m).

255. FISH-EATING STAR

Stylasterias forreri

Identification: The very long arms have sharp spines and abundant clusters of pedicellariae. Color varies from black to brown. *Size:* Diameter to about 20 inches (50 cm). *Range and Habitat:* Southern Alaska to San Diego, California. On rocky as well as soft bottoms; in subtidal depths from 20 to about 180 ft (6 to 540 m). *Natural History:* The pedicellariae are used to capture small fishes such as sculpins (Cottidae). These cryptic sea stars also feed on molluscs.

256. SUNFLOWER STAR

Pycnopodia helianthoides

Identification: This large sea star usually has 20 to 24 flexible arms. Juveniles have five arms. The color varies from purple to brown, orange, or yellow. *Size:* Diameter to about 36 inches (90 cm). *Range and Habitat:* Aleutian Islands to Isla Todos Santos, Baja California. On rocky as well as soft bottoms; from the low intertidal to about 1450 ft (435 m). *Natural History:* In Alaskan waters, the king crab (# 175) feeds on this sea star.

257. FRAGILE STAR

Linckia columbiae

Identification: Each arm of this star is almost round in cross section. Arms rarely of equal length. The number of arms varies from one to nine. *Size:* Diameter to about 4 inches (100 mm). *Range and Habitat:* Southern California to Colombia and Galapagos Islands. On rocks; from the low intertidal to about 240 ft (72 m). *Natural History:* This sea star often breaks off its arms (autotomize) when threatened and as a means of asexual reproduction.

258. CHANNELED SEA STAR

Tethyaster canaliculatus

Identification: This sea star has five, flattened, tapering arms, usually unequal in length. Color is orange-red. *Size:* Diameter to about 20 inches (50 cm). *Range and Habitat:* Coronado Islands, Baja California to Gulf of California and Panama. On soft bottoms, from 20 to 585 ft (6 to 78 m).

259. GUNPOWDER STAR
Gephyreaster swifti

Identification: This five-rayed star has large marginal plates on the edges of each arm. The color ranges from preyish pink to light orange. *Size*: Diameter to about 16 inches (40 cm). *Range and Habitat*: Sea of Japan to Bering Sea and south to Oregon. On mud, sand, gravel and rock, most often on soft bottoms. Depth range 50 ft. (15 m) to 8500 ft. (250 m). *Natural History*: Gunpowder stars feed on snails, brittle stars and small crustaceans, as well as sediment.

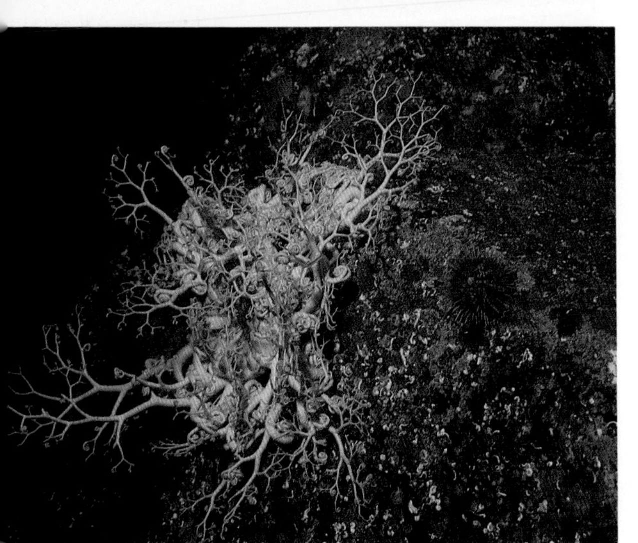

CLASS OPHIUROIDEA
Basket, Serpent, Brittle Stars

260. COMMON BASKET STAR
Gorgonocephalus eucnemis

Identification: This distinctive animal has arms with many branches and branchlets. Color varies from almost white to orange-red, pink, tan, and beige. *Size:* Diameter to about 30 inches (76 cm). *Range and Habitat:* Circumpolar. On this coast from Bering Sea to southern California. On rocks; in subtidal depths to about 500 ft (150 m).

Marc Chamberlain

261. BANDED SERPENT STAR
Ophioderma panamensis

Identification: These large brittle stars have black to dark brown arms, with light colored rings, and a dark star-shaped blotch on the body. *Size*: Diameter to about 12 inches (30 cm). Range and Habitat: Southern California and Baja California on soft bottoms, intertidal to 90 ft (27 m).

262. SPINY BRITTLE STAR

Ophiothrix spiculata

Identification: This brittle star is easily identified by the long, erect spines on the arms and disc. The spines have thornlike spinlets. Color varies. *Size*: Diameter to about 15 inches (38 cm). *Range and Habitat:* Central California to Galapagos Islands and Peru. On rocky and soft bottoms; from the low intertidal to about 6600 ft (2000 m). *Natural History:* This brittle star occurs in very large numbers in some areas. This star is becoming a dominent species with million of individuals covering substrates.

263. BRITTLE STAR

Ophiopteris papillosa

Identification: The arms of this spiny brittle star are about 3 to 4.5 times as long as the disc diameter. The long, flat spines are blunt. Color is chocolate-brown, with bands on the arms. *Size:* Diameter to about 7 inches (175 mm). *Range and Habitat:* Barkley Sound, British Columbia, to Isla Cedros, Baja California. On rocks and kelp holdfasts; from the low intertidal to about 460 ft (140 m).

264. BRITTLE STAR

Ophiopsilla californica

Identification: The long, slender arms are covered with long, slender, sharp spines. This brittle star remains buried in sand or mud bottoms with only the arms exposed. *Size:* Diameter to 12 inches (300 mm). *Range and Habitat:* Central California. In soft bottoms; shallow subtidal depths.

265. SMOOTH BRITTLE STAR

Ophioplocus esmarki

Identification: A relatively smooth brittle star with a large disc and short spines that can be folded against the arms. Color brown to gray-brown. *Size:* Diameter to about 6 inches (150 mm). *Range and Habitat:* Tomales Bay to San Diego, California. On soft bottoms; from the low intertidal to about 230 ft (70 m). *Natural History:* A slow moving brittle star that broods its eggs. The eggs develop directly into little brittle stars.

PHYLUM CHORDATA

SUBPHYLUM UROCHORDATA

Tunicates, Sea Squirts

These animals, strangely enough, are the closest relatives to man. For the most part, the adults are sessile animals, attached to the substrate; however, many are pelagic. The characters that we and all vertebrates share with the tunicates include: a dorsal nerve cord that begins in the anterior end of the animal as a brain, a skeletal rod located underneath, which supports the nerve chord, called the notochord, and a pharnyx that develops gill slits. Some of the characters only appear in the embryonic or larval stages. Tunicates can exist as solitary individuals, as an aggregation, or in colonies. The typical tunicate has an incurrent syphon for the intake of water and oxygen bearing food and an excurrent syphon for discharging the water with body wastes. All of the organ systems of higher vertebrates are present with the exception of specialized sensory organs such as eyes.

Tunicates are filter feeders and usually live in areas where there are currents with entrained with food organisms. All species practice sexual reproduction, however, colonial tunicates can multiply by cloning from a single sexually produced individual. Tunicates are preyed on by invertebrates and fishes. They have been used in research by embryologists and cell biologists. Taxonomists have recorded about 90 species from Alaska to California.

266. LOBED TUNICATE

Cystodytes lobatus

Identification: Young colonies form flat sheets, but older colonies form large masses with convolutions or ridges. Color varies from gray, whitish, orange-pink, or lavender. *Size:* Diameter of adult colony about 10 inches (250 mm) and height to about 4 inches (100 mm). *Range and Habitat:* Vancouver Island, British Columbia to Isla San Geronimo, Baja California. On rocks; from the very low intertidal to about 666 ft (200 m).

267. LIGHT BULB TUNICATE

Clavelina huntsmani

Identification: The zooids of this social tunicate have a transparent tunic and visible internal organs that look like the filament of a light bulb. *Size:* Height of zooids to about 2 inches (50 mm). Diameter of colony 20 inches (50 cm). *Range and Habitat:* British Columbia to Punta Banda, Baja California. On rocks; from the low intertidal to about 100 ft (30m).

268. COLONIAL TUNICATE

Distaplia occidentalis

Identification: The colonies have mushroom-shaped lobes. Color varies greatly. *Size:* Diameter of colony to about 4 inches (100 mm). *Range and Habitat:* Vancouver Island, British Columbia to San Diego, California. Very common in British Columbia. On rocks; from the low intertidal to about 50 ft (15 m).

269. SOCIAL TUNICATE

Pycnoclavella stanleyi

Identification: The exposed, upright orange or gold striped projections are branchial baskets of expanded zooids. Each basket represents a single individual. *Size*: Individual lengths to about 0.75 inches (20 mm). *Range and Habitat:* Vancouver Island, British Columbia to Isla San Geronimo, Baja, California. On rocks; from the very low intertidal to about 30 ft (10 m).

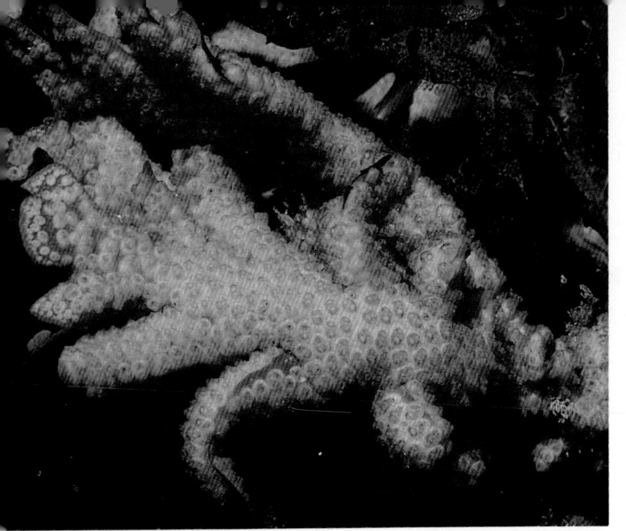

270. COLONIAL TUNICATE

Metandrocarpa dura

Identification: Similar to *M. taylori* (not included here) but individuals are packed together in thick sheets that envelop seaweed stems. Color ranges from bright red to orange. *Size:* individuals about 1/8 to 1/4 inches (3-6 mm) in diameter. *Range and Habitat:* British Columbia to Baja California, attached to brown algae. Subtidal to about 50 ft. (15 m).

271. ELEPHANT EAR TUNICATE

Polyclinum planum

Identification: The ear-like, lobed colony of zooids is attached to the substrate by a slender stalk. Color is brownish. *Size:* Diameter of lobe to about 8 inches (200 mm). *Range and Habitat:* Fort Bragg, California, to Isla San Geronimo, Baja California. On rocks; from the very low intertidal to about 100 ft (30 m).

272. COLONIAL TUNICATE

Didemnum carnulentum

Identification: The exposed portions of the colony are smooth, with many small openings and one large aperture. Color is white or gray. *Size:* Diameter of colonies to about 5 inches (125 mm). *Range and Habitat:* Oregon to Panama. On rocks; from the intertidal to about 100 ft (30 m).

273. SOLITARY TUNICATE

Corella willmeriana

Identification: The translucent tunic of this upright tunicate is distinctive. It is large, bulbous and transparent, with internal organs visible. *Size:* Height to about 3 inches (75 mm). *Range and Habitat:* Southern Alaska to Pt. Fermin, California On rocks; in subtidal depths to about 250 ft (75 m).

274. GLASSY TUNICATE

Ascidia paratropa

Identification: This large, solitary tunicate has a prominent tubercle on the sides and two siphons at the tip. The siphons are unequal in size. The excurrent siphon is the larger. *Size*: Height to about 6 inches (150 mm). *Range and Habitat:* Ungo Strait, Aleutian Islands, Alaska, to southern Monterey County, California. On rocky substrate; from the very low intertidal (rare) to depths of about 330 ft (100 m).

275. COLONIAL TUNICATE

***Botrylloides* sp.**

Identification: This unidentified colonial tunicate forms globular masses. There is at least one large aperture on the upperside. *Size:* Diameter of colony to about 3 inches (75 mm). *Range and Habitat:* Southern California. On rocks in the shallow subtidal.

276. SOLITARY TUNICATE

Cnemidocarpa finmarkiensis

Identification: This hemispherical solitary tunicate has a smooth, opaque tunic. The two siphons on the upper surface are about equal in size. Color is usually bright red. *Size:* Diameter of base to about 2 inches (50 mm). *Range and Habitat:* Circumpolar. On this coast from Alaska to Point Conception, California. On rocks; from the very low intertidal to about 160 ft (50 m). *Natural History:* This distinctive animal is preyed on by the sea star *Orthasterias koehleri* (# 251).

277. COLONIAL TUNICATE

Metandrocarpa taylori

Identification: Each individual zooid of the colony has two siphons. The zooids are joined at their bases by a sheet of tunic which encrusts the substrate. Color is usually red or orange, rarely green or yellow. *Size:* Diameter of colonies to about 8 inches (200 mm). *Range and Habitat:* British Columbia to San Diego, California. On rocks; from the low intertidal to about 65 ft (20 m).

278. STALKED TUNICATE

Styela montereyensis

Identification: The long stalk of this solitary tunicate is distinctive. The elongated tunic has longitudinal ridges. Color is usually yellow to dark red-brown. *Size*: Height to about 10 inches (250 mm). *Range and Habitat:* Vancouver Island, British Columbia to Isla San Geronimo, Baja California. Attached to rocks; from the low intertidal to about 100 ft (300 m).

279. PEANUT SEA SQUIRT

Styela gibbsii

Identification: These solitary tunicates usually occur in groups. Their peanut shape, short stalk and tan to orange-brown color are distinctive. *Size*: Height to about 6 inches (15 cm). *Range and Habitat*: British Columbia to southern California. On rocks, low intertidal and shallow subtidal.

280. SPINY-HEADED TUNICATE

Boltenia villosa

Identification: This stalked tunicate has a small, rounded body covered with many spines. The stalk is about twice as long as the diameter of the body. Color is orange-brown. *Size:* Diameter of body to about 1 inch (25 mm). Height to about 2.5 inches (60 mm). *Range and Habitat:* Southern Alaska to San Diego, California. Attached to rocks and other hard substrate; from the very low intertidal to about 330 ft (100 m). *Natural History: Boltenia* is preyed on by some sea stars.

281. SEA PEACH

Halocynthia aurantia

Identification: This distinctive, solitary, upright, smooth tunicate has two uneven sized siphons on its upper end. The color ranges from yellowish to orange. *Size:* Height to about 6 inches (150 mm). *Range and Habitat:* Chukchi Sea to Puget Sound, Washington.On rocks; in subtidal depths to about 330 ft (100 m).

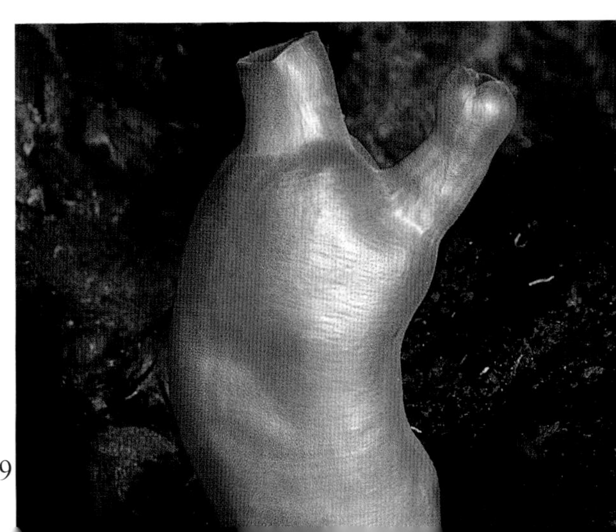

282. SPINY SEA SQUIRT

Halocynthia hilgendorfi igoboja

Identification: The globular body of this solitary, spiny tunicate is opaque, but because the spines tend to collect mud and detritus, the opaqueness is not apparent. The siphons are orange-red when open. *Size*: Diameter to about 4 inches (10 cm). *Range and Habitat*: British Columbia to Channel Islands, California. On rocks and gravel, from the intertidal to about 540 feet (165 m).

283. SOLITARY TUNICATE

Pyura haustor

Identification: The globular-shaped body of this solitary tunicate is usually hidden by other encrusting organisms. The exposed siphons have a distinctive red-rimmed opening. *Size:* Diameter to about 3 inches (75 mm). Height to about 3 inches (75 mm). *Range and Habitat:* Shumagin Islands, Alaska to San Diego, California. Attached to hard substrate; from the low intertidal to about 660 ft (200 m). *Natural History:* The sea star, *Solaster stimpsoni* (# 242), preys on this tunicate.

284. COLONIAL TUNICATE

Eunerdmania claviformis

Identification: The very elongated, club-shaped zooids are almost translucent. *Size:* Zooid height to about 2 inches (50 mm). *Range and Habitat:* British Columbia to San Diego, California. On rocks; from the low intertidal to shallow subtidal depths.

285. COLONIAL TUNICATE

Archidistoma molle

Identification: The distinctive colonies are in the form of smooth round or oval lobes. Color opaque white with red-tipped zooids. *Size*: Diameter to about 4 inches (10 cm). *Range and Habitat*: British Columbia to Carmel Bay, California. On rocks, in beds of surf grass, low intertidal to about 35 ft (10 m).

CLASS THALIACEA

Pelagic Tunicates

286. PELAGIC TUNICATE

Thetys vagina

Identification: A tunicate with a translucent tunic, with internal organs readily visible. Occurs as solitary (sexual form) individual and in chains of attached animals (asexual budding form). *Size:* Length to about 10 inches (25 mm). *Range and Habitat:* Found in most of the world's oceans, particularly in temperate waters. Pelagic. *Natural History:* These pelagic invertebrates are important food items for some species of rockfish, particularly the blue rockfish, *Sebastes mystinus,* and the mola, *Mola mola.*

Solitary form

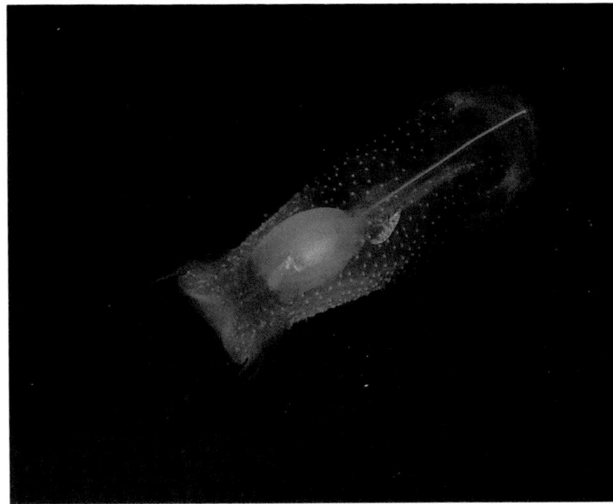

Clonal form

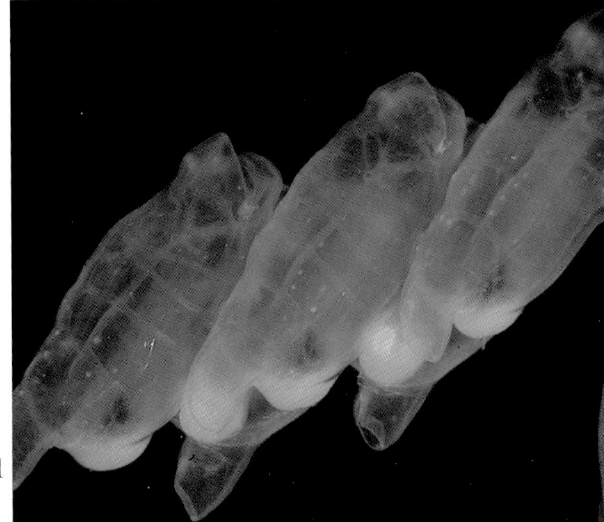

SELECTED REFERENCES

Austin, W.C. 1985. *An annotated checklist of marine invertebrates in the cold temperate northeast Pacific.* Khoyatan Marine Laboratory, Cowichan Bay, B.C. 3 Vol., 682 p.

Barr, L., and N. Barr. 1983. *Under Alaskan Seas: The shallow water marine invertebrates.* Alaska Northwest Publishing Co. 208 p.

Behrens, D.W. 1991. *Pacific Coast Nudibranchs - A Guide to the Opisthobranchs Alaska to Baja California.* Sea Challengers, Monterey, CA. 107 p.

Cairns, S.D. 1991. *Common and scientific names of aquatic invertebrates from the United States and Canada: Cnidaria and Ctenophora.* Am. Fish. Soc. Spec. Pub. 22. v 75 pp.

Jensen, G.C. 1995. *Pacific Coast Crabs and Shrimps*, Sea Challengers, Monterey, CA 87 p.

Harbo, R.M. 1999. *Whelks to Whales Coastal Marine Life of the Pacific Northwest.* Harbour Publishing. Madeira Park, British Columbia. 245 p.

Kozloff, E.N. 1983. *Seashore life of the northern Pacific* coast, *an illustrated guide to northern California, Oregon, Washington, and British Columbia.* Univ. Wash. Press. 370 p.

Lambert, P. 1945. *The sea stars of British Columbia.* British Columbia Prov. Mus. Handbook 39. 153 p.

McLean, J.H. 1969. *Marine shells of southern California.* Los Angeles Co. Mus. of Nat. Hist., Expostion Park, Los Angeles, CA. Sci. Ser. 24, Zool. No. 11. 104 p.

Morris, P.A. 1966. *A field guide to shells of the Pacific coast and Hawaii.* Houghton Mifflin Co., Boston, MA 297 p.

Morris, R.H., D.P. Abbott, and E.G. Haderlie, 1980. *Intertidal Invertebrates of California.* Stanford Univ. Press, Stanford, CA. 690 p.

Ricketts, E.F., J. Calvin, J.W. Hedgpeth. 1985. *Between Pacific Tides.* 5th Ed. Revised by David W. Phillips. Stanford Univ. Press. 652 p.

Smith, R.I., and J.T. Carlton (editors), 1975. *Light's Manual: Intertidal Invertebrates of the Central California Coast.* Third ed. Univ. of Calif. Press, Berkeley. 716 p.

Stachowitsch, M. 1992. *The Invertebrates: an Illustrated Glossary.* John Wiley and Sons, Inc. NY. 676 p.

Turgeon, D.D. 1988. *Common and scientific names of aquatic invertebrates from the United States and Canada: Mollusks.* Amer. Fish. Soc. Spec. Pub. 16. 277 p.

Williams, A.B., L.G. Abele, D.L. Felder, H.H. Hobbs, Jr. R.B. Manning, P.A. McLaughlin, and I.P. Farfante. 1989. *Common and scientific names of aquatic invertebrates from the United States and Canada: Decapod crustaceans.* American Fisheries Soc. Spec. Pub. 17. 77 p.

Wrobel, D. and C. Mills. 1998. *Pacific Coast Pelagic Invertebrates - A Guide to the Common Gelatinous Animals.* Sea Challengers, Monterey, CA. 112 p.

INDEX TO COMMON NAMES

INDEX TO SCIENTIFIC NAMES

INDEX TO SCIENTIFIC NAMES (cont.)

INDEX TO SCIENTIFIC NAMES (cont.)